RAIN drizzle & FOG

RAIN drizzle & FOG

NEWFOUNDLAND WEATHER STORIES

BY

Sheilah Roberts

INTRODUCTION BY

Ryan Snoddon

BOULDER
PUBLICATIONS

Library and Archives Canada Cataloguing in Publication

Roberts, Sheilah, 1954-, author
    Rain, Drizzle, and Fog : Newfoundland weather stories
    / Sheilah Roberts.

ISBN 978-1-927099-57-5 (pbk.)

    1. Newfoundland and Labrador--Climate.   I. Title.

QC985.5.N5R62 2014   551.69718   C2014-904569-7

PUBLISHED BY
Boulder Publications
Portugal Cove-St. Philip's, Newfoundland and Labrador
www.boulderpublications.ca

© 2014 Sheilah Roberts

EDITOR
Stephanie Porter

COPY EDITOR
Iona Bulgin

COVER ILLUSTRATION & DESIGN
Jud Haynes
www.judhaynes.com

INTERIOR LAYOUT & DESIGN
Mike Mouland with Jud Haynes

Printed in Canada

We acknowledge the financial support of the Government of Newfoundland and Labrador through
the Department of Tourism, Culture and Recreation.

We acknowledge the financial support for our publishing program by the Government of Canada,
including the Canada Council for the Arts, and the Department of Canadian Heritage through the
Canada Book Fund.

CONTENTS

GRAND BANK FISHING SCHOONERS AT ANCHOR IN FOG, ST. JOHN'S HARBOUR, 1929

The Rooms Provincial Archives Division, VA 41-8 / Wallace Robinson MacAskill

Always interesting and unpredictable (even for the
weatherman), the weather here in Newfoundland and Labrador is
truly the number one topic of conversation. Even during a stretch
of nice weather, everyone talks about how they can't believe that
it's so pleasant outside. The weather is a part of who we are; it
always has been and it always will be. We live it, we breathe it.
When the weather is nice and the sun is shining, no one can argue
that this province is one of the most beautiful places on the
planet. However, as we know, the weather here can change in a
hurry. A quick shift in the wind and we see some of the most
powerful and destructive weather Mother Nature has to offer.

I've quickly learned that Newfoundlanders and Labradorians
have a love-hate relationship with the weather. Let's be honest:
four snowstorms in one week is enough to send even the most
passionate winter enthusiast over the deep end, with shovel in
hand. However, it seems with the accumulation of each centimetre
of snow through the never-ending winter, comes a little more pride
in our collective hardiness. It helps us appreciate the next sunny
day even more.

In this province we know we can tough it out no matter what
the weather. Winters like the infamous season of 2000–01, when
more than 6 metres (21 feet) of the white stuff fell in St. John's,
and summers like 2011 (complete with "Juneuary" and "Fogust")

only make us stronger. These only-in-Newfoundland-and-Labrador stories seem to give us some kind of twisted bragging rights that the weather here is crazier than anywhere else in the country, or maybe even the world.

As miserable as it might be at times, Newfoundlanders and Labradorians have had no choice but to embrace, or at the very least endure, the weather. For generations, people's livelihoods have been made outdoors and on the water, where the weather is often the most harsh and punishing. Our location almost guarantees we'll see the unsettled and the unexpected when it comes to the weather. We sit smack dab in the middle of an air mass battle zone. We have cold dry Arctic air to our north-west, cold moist maritime air to our north and east, and warmer moist air to our southeast, thanks to the Gulf Stream. Large temperature contrasts and moisture are the fuel for the biggest storms on the planet, and we have all these ingredients at the ready.

Over the years, this province has experienced every type of weather event imaginable. Hurricanes, tropical storms, paralyzing winter blizzards, avalanches, crippling ice storms, relentless snow squalls, 200-plus-kilometre-per-hour windstorms, weeks of relentless thick-as-pea-soup fog, massive moisture-laden fall Nor'easters, tornadoes, and even droughts and wildfires have affected residents over the past 400 years. Everything Mother Nature has to offer is on the menu.

Our weather history is long and varied. As you'll read in the pages of *Rain, Drizzle & Fog*, even back as far as the 1600s the earliest settlers were tested by the elements and we've been battling them ever since. Hurricane Igor, the Badger floods, the Great Drought of 1911, and the Great Hurricane of 1775: they are forever a part of our history. Those storms and events are a constant reminder of where we live and how vulnerable we really are.

This is a book for weather weenies, history buffs, and everyone in between. The stories and first-hand accounts of those ferocious and famous storms of the past are truly fascinating. From newspaper clippings to weather journals, Sheilah has put together a weather history book, which is the first of its kind in Newfoundland and Labrador. Add in the weather science explainers, facts, folklore, and, of course, the helpful glossary, and what you really have in your hands is the weather bible for Newfoundland and Labrador—a truly great read. I hope you enjoy it as much as I did.

Ryan Snoddon

METEOROLOGIST, CBC NEWFOUNDLAND & LABRADOR

DIGGING OUT A FIRE HYDRANT IN 5.5 METRES OF SNOW IN ST. JOHN'S, CA. 1920
City of St. John's Archives

PREFACE

Disasters and major weather events fill the annals of
Newfoundland history. This book resulted from an idea that it
would be interesting to consider the weather in this province
over the span of several hundred years. Weather events are
arranged first by season, and then chronologically by month and
year—for example, the January chapter has weather stories from
1611 to 1977. These are interspersed with notes on weather science
and traditional weather lore. At the end of the book is a glossary
of Newfoundland weather terms. Because early sources contained
different measurement scales than those used in 2014, some
conversion formulas are also included.

As I could only choose a few events for each month, I apologize
if I have missed your favourite weather story.

WHAT WE CALL WEATHER

Newfoundland and Labrador is well known for its erratic weather. Some days residents can experience all four seasons in a 24-hour period. You might look out the window one morning and see the beginnings of a gorgeous day—but look closer and you'll notice something grey creeping up the harbour. Fog can quickly turn a balmy summer day into a cold, damp monster, when the only joy to be gleaned is a fire in the grate to chase off the chill. Before you know it, the weather changes again, and the rain begins, drumming on the windowpane, beating out a rhythm accompanied by a trumpet of wind—just another day in St. John's, Burin, Twillingate, St. Anthony, or L'Anse au Clair.

All Atlantic Canadians are familiar with the infamous trios of rain, drizzle, and fog and, of course, wind, snow, and sleet. As American humorist Kin Hubbard (1868–1930) once wrote, "Don't knock the weather. If it didn't change once in a while, nine out of ten people couldn't start a conversation."[1]

Newfoundlanders and Labradorians love to talk about the weather. They depend on the weather report for planning not only their work but also their recreation. On the island, Environment Canada has three automated telephone answering devices, in St. John's, Gander, and Corner Brook, with recordings of the latest weather reports. In August 2010, they received, on average, 6,900 calls per day.[2] Online weather forecasts are also popular, and in May 2013, for example, Environment Canada's website recorded 39,348,649 visits from across Canada, of which 1,127,076 were from Newfoundland.[3]

Newfoundlanders employ a rich vocabulary for describing the weather they experience—civil, mauzy, muggy, misky, loggy, scuddy. What is a screecher? A fairy squall? A norther? (Check the list of Newfoundland weather terms at the back of the book.) Often unpleasant weather is simply labelled *weather*: "Looks like we're going to have some weather."

Known as one of the stormiest parts of North America, Newfoundland has an impressive history of destructive storms. As Environment Canada climatologist David Phillips says: "There's nothing wimpy about the climate here. Of all the major Canadian cities, St. John's is the foggiest, snowiest, wettest, windiest, and cloudiest. It also has more days with freezing rain and wet weather than any other city [in Canada]."[4]

A mild phase in the world's climate was recorded in the early medieval period (800–1300). When the ice in the northern seas melted, it allowed the Norsemen to voyage to Greenland, Iceland, and the Northern Peninsula of Newfoundland. After this period of warming came a mini ice age, which occurred, with the exception of a few short mild periods, between 1550 and 1850.[5] During the cooling period in sixteenth- and seventeenth-century Europe, severely cold winters and cool summers were typical. In London, the Thames froze solid and "frost fairs" were held on the ice.

Probably in a desperate measure to warm themselves, English adventurers began to look farther afield. In the sixteenth and seventeenth centuries, it was a commonly held belief that climate was constant at any latitude. Looking westward, they calculated that the land that would later become Jamestown, Virginia, was situated at about the same latitude as southern Spain. They envisioned the dry, hot climate of the Mediterranean and a landscape dotted with olives and fiery peppers. When their eyes turned to the north and the fish-rich waters of Newfoundland, they saw it was on the same latitude as Paris, France, with its gentle winters and romantic springtimes.

When settlers arrived on the southeastern Atlantic coast in 1607, and started the settlement that later became Jamestown, they were probably disappointed in the weather. But those that arrived in Cupers Cove (Cupids), Newfoundland, a few years later, must have been disheartened. The weather was not at all like that of Paris. The eastern coast of North America was much colder, wetter, and foggier than they had expected. No romance there, especially not in Newfoundland's spring—not many flowers, and the only gifts of nature were the icebergs that floated down the coast in late spring and early summer.

Any hopes of productive agricultural activities on the island were short-lived. In addition to its unfavourable weather, Newfoundland also offered soil that was not fertile. The early settlers, always ones to look on the bright side, no doubt offered themselves this cheerful advice: "hee is wretched who believes himself wretched."[6] The colonists soldiered on. Even though Newfoundland soil did not produce the array of crops the settlers hoped for, the optimists in the group noted that, because of the cold, Newfoundland and the American colonies would be a wonderful future market for English woollen cloth and that this would put money in homeland coffers.[7]

Making sense of the Newfoundland environment in these early days of settlement was a matter of survival. To avoid starvation, the settlers had to know what crops they could grow and, to do that, they needed to know what type of weather to expect. In 1610, John Guy started the colony at Cupers Cove in present-day Conception Bay North on the Avalon Peninsula. Henry Crout, one of the settlers and Sir Percival Willoughby's first agent in Newfoundland, recorded the weather in his journals and also wrote about it in his letters home. John Mason, governor of Cupers Cove from 1615 to 1621, writing home to England, painted a pleasant picture of Newfoundland summer: "The aire subtle & wholesome the Summer season pleasant conforme to the like latitude in Europe ..." The winters soon surprised him, as they were colder than those at the same latitude in Europe. He described the winter as being like that in the northern parts of Scotland or the Hebrides but with the advantage of longer days (England lies farther north).[8] Mason did not stay in Newfoundland. Perhaps seeking warmer climes, he moved south in 1629 to start the colony which later became New Hampshire.

Sir Richard Whitbourne's connection with Newfoundland goes back before Mason's, to 1579 when he came from England as a 15-year-old. Almost 40 years later, in 1618, he was appointed governor of the Welsh colony New Cambriol. Founded by Sir William Vaughan, a Welsh poet and adventurer, New Cambriol was located on the southern tip of the Avalon Peninsula on a parcel of land that stretched from present-day Calvert across to Placentia and down to the southern tip of Trepassey. When Whitbourne took over the colony of New Cambriol, it was in a lamentable state and, despite his best efforts, it eventually failed. Whitbourne, however, did his best to promote colonization in Newfoundland, and he was not reticent about embellishing certain aspects of the climate. And no wonder: he'd worked himself into an enviable position in the new land. The clergy were instructed by their archbishops through the Privy Council to distribute Whitbourne's book *A discourse and discovery of New-found-land* throughout England and to collect "voluntary contributions" to reimburse him for his work. Whitbourne did not exactly lie about the colonies, but he did downplay their negative aspects. He ridiculed those who found the Newfoundland climate uncomfortable, accusing complainers of being soft, accustomed to sitting in taverns by the fire, or having been "touched with the French disease."[9]

In a letter to England on August 17, 1622, Captain Edward Wynne, the governor of Cupers Cove, wrote of what likely was a mild winter: "Neither was it so colde here the last Winter as in England the yeare before. I remember but three severall days of hard weather indeed, and they not extreame neyther: for I have knowne greater Frosts, and farre greater Snows in our owne countrey."[10]

STARS IN THE NORTH AMERICAN SKY
Books of Knowledge, 1912

In 1621, Sir George Calvert had sent men from England to start a colony in Ferryland. He did not arrive until 1627, staying through the reasonably mild winter of 1627–28 and then returning to England. When he came back to Ferryland in 1628 with his family, however, he experienced "intolerable cold."[11] It did not help that the colonists had chosen to build the Colony of Avalon (Ferryland) on an exposed point of land.

When the earliest arrivals discovered that the weather was not what they had expected, they tried to come up with reasons why it did not. One imaginative explanation came from Samuel Purchas, an English writer at the turn of the seventeenth century, who collected and edited information about travelling the world; in his book about the "new found lands" he noted that the stars were smaller and fewer in the North American sky, which therefore made its climate colder: "This wont of the Sunne and Starres is one cause of greater cold in those parts [North America] than in these [southern climes]."[12]

On the other hand, John Mason of Cupers Cove blamed the cold on the many lakes and ponds in Newfoundland, since fresh waters were "more naturally cold than salt and both colder than the earth."[13] Newfoundland was cooler than England, he suggested, because there were not as many people or houses, and, therefore, fewer house fires warmed the air: "the Country slenderly peopled, voide of Townes and Cities, whereof Europe is full; the smoake whereof and heate of fires much qualifieth the coldnesse of the Aire."[14]

Richard Whitbourne devised a scheme for raising these cool temperatures on the island. He reasoned that cutting down the woods would clear the vapours and allow the sun's rays to penetrate and warm the atmosphere:

> Hot beames of the Sunne might pearce into the earth and
> stones there, so speedily as it doth in some other Countreys,
> that lye vnder the same elevation of the Pole, it would then

there make such a reflection of heate, that it would much lessen these Fogs, and also make the Countrey much hotter Winter and Summer, and thereby the earth will bud forth her blossoms and fruites more timely in the yeere, then now it doth, ... and thereby those Ilands of Ice that come on that coast at any time, will the sooner dissolve.[15]

Those "Ilands of Ice" that clogged up the ocean and bays were blamed, as they are in the twenty-first century, for making the air cold. Newfoundlanders and Labradorians always look for ways to explain the province's weather. But even with sophisticated meteorological instruments and equipment, it is difficult to forecast the weather with complete accuracy.

In the mid-latitudes, winds generally blow from west to east. Because land heats up and cools off much more quickly than water, the two climates—the east coast of North America and the west coast of Europe—differ widely. On the east coast of North America, weather is mainly affected by air masses that have travelled over the land from the west. As a result, eastern North America has a temperate continental climate featuring greater extremes of heat and cold,[16] with frigid winter and hot summers, buffered by a short spring and fall. In England and on the west coast of the European continent, the climate is influenced by air masses moving over water. The ocean air picks up and releases heat slowly, creating a delay in the onset of seasons, along with milder winters and cooler summers.[17]

The island of Newfoundland has a combination of climates. Situated in the Atlantic Ocean and influenced by the water around it, its coastal climate is predominantly maritime, while that of its interior can be considered closer to a continental climate. In Labrador, the climate is boreal (sometimes considered subarctic), with cold winters and one to three months with average temperatures of 10°C or higher. Labrador lies in the path of air flowing off mainland North America, which gives its climate more seasonal contrasts than the maritime climate of the island. Labrador's climate becomes increasingly cool in higher latitudes, transitioning to polar conditions at its northern extremes.[18]

Other influences on Newfoundland's climate include the cold Labrador Current and the warm Gulf Stream. The latter begins upstream from Cape Hatteras and comes up from the Gulf of Mexico to the Strait of Florida, passing along the eastern seaboard on its way to Newfoundland. It is one of the world's fastest moving currents at 30 million cubic metres per second, increasing to 150 million cubic metres per second off the coast of Newfoundland.[19] Benjamin Franklin discovered that the powerful Gulf Stream, which can now be seen from space, hindered the speed of British mail boats on their journey across the Atlantic Ocean to North America. Any ship that followed it going westward against the current made slow progress.[20] The Gulf Stream contributes to the development of weather systems that affect Newfoundland and Labrador; the temperature contrast between the cold air and the warm waters of the Gulf Stream can provide energy to developing storms.

The Labrador Current flows down from the Arctic Ocean, passes along the coast of Labrador, and meets the Gulf Stream just off the Grand Banks. The cold current keeps the water cool, so that when warm air blows over it, the moisture condenses and forms fog. Stephen Parmenius, a Hungarian in charge of recording the travels of Sir Humphrey Gilbert on the *Swallow* in 1583, described the Newfoundland fog this way:

> the ayre upon land is indifferent cleare, but at Sea towardes the East there is nothing els but perpetual mistes and in the Sea it selfe, about the Banke (for so they call the place where they find ground at fourty leagues distant from the shoare, and where they began to fishe) there is no day without rayne.[21]

Many Newfoundlanders do practically live in fog. Argentia, on the Burin Peninsula, has more fog than almost anywhere else in Canada. In 1966, the community recorded 230 days of fog.[22] That does not leave much room for sunshine.

Statistics on which city is the foggiest, snowiest, rainiest, or windiest changes constantly, but St. John's, Newfoundland, has been named the fog capital of Canada, with an average of 121 foggy days per year.[23] As for the foggiest point of land in the world, Cape Race, wins hands-down with 3,792 hours or 158 full days of fog a year.[24]

No wonder Newfoundlanders need vitamin D supplements.

Blow, blow, thou winter wind
Thou art not so unkind,
As man's ingratitude.

WILLIAM SHAKESPEARE

ST. JOHN'S IN THE SNOWY LONG AGO NEW GOWER STREET, LOOKING EASTWARD, CA. 1925

Archives and Special Collections Division, Queen Elizabeth II Library, Memorial University of Newfoundland (ASCD)

JANUARY

TRADITIONAL WEATHER LORE

Before the advent of radio and television weather reports, Newfoundlanders depended on the weather lore passed down from their forefathers. Take snow, for example. How often has this adage—*Big snow, little snow. Little snow, big snow.*—been true? Large flakes of snow result in little accumulation, but if the flakes are small, be prepared for a heavy snowfall.[25] One explanation: Colder, drier snow is often low in density. Comprised of delicate crystals, it can build into a deep snowfall. If snow forms in warmer conditions, it tends to be of higher density and becomes wet and clumpy; therefore, this "big snow" can be shallower, slushier, and more compact, making it look like a smaller snowfall. As well, the largest snowflakes often form in the strong updrafts associated with unstable air masses, such as when cold air moves over relatively warm water. This often gives rise to heavy bursts of snow comprised of big flakes, but, like a quick summer shower, they do not last long. Smaller snowflakes are often found in large weather systems, where the updrafts are gentler but snowfalls tend to last for a much longer time, leading to considerable accumulations.[26]

1611

Cupers Cove, Newfoundland, was the first English settlement
in Canada and the second in North America after Jamestown,
Virginia. The first winter that settlers stayed at Cupers Cove was
a mild one, and John Guy, governor of the colony in 1610 and
1611, wrote this description in a letter he sent back to England.
He concluded that winter in Newfoundland was milder than
that at home:

> The most part of January and February unto the middle of
> March the frost continued: the winde being for the most
> part Westerly, and now and then Northerly: notwithstanding
> three or four times, when the winde was at South, it began
> to thaw and did raine. That which fell in this season was
> for the most part Snow, which with the heate of the Sunne
> would be consumed in the open places with a few dayes.
> That which abode longest was in February. During this time
> many dayes the Sun shone warme and bright from morning
> to night. Notwithstanding the length of this frosty weather,
> small brookes that did run almost in level with a slow course
> were not the whole winter three nights over frozen so thicke
> as that the Ice could beare a Dogge to goe over it ... [this] I
> found by good proofe for every morning I went to the brooke
> which runneth by our house to wash.[27]

Guy returned to England in 1611, leaving his brother-in-law,
William Colston, in charge of the colony until his return, in 1612.
Colston kept a diary, which included descriptions of the weather.

His report for the winter of 1611–12 was less optimistic than
Guy's description. Colston's diary (which Samuel Purchas called
"very tedious") did not survive, but Purchas wrote a summary of
Colston's report on the winter weather that year: "It appeareth
that the weather was somewhat more intemperate then it had
beene the yeare before, but not intolerable, nor perhaps so bad
as we have it sometims in England."[28]

1697

The French and English settled peacefully on the island of
Newfoundland in the early 1600s. Their counterparts overseas were
not so amicable, and conflicts between the two countries often
spilled over into Newfoundland. By the mid-1600s, French ships
fishing off the coast of Newfoundland outnumbered the English
roughly two to one,[29] so it was only logical that the French would
try to gain a stronger foothold on the island. In 1696, Montreal-
born Pierre Le Moyne d'Iberville left France in the spring of
1696 with orders to attack English stations along the Atlantic
coast. He arrived in Placentia (Plaisance), the French capital of
Newfoundland, on September 11. As he and his men walked across
the Avalon Peninsula that fall and winter, they successfully attacked
English settlements by striking from inland, thereby avoiding the
heavy gun batteries pointing out to sea. He raided and burned 36
communities on the east coast of the Avalon Peninsula, including
Ferryland, Bay Bulls, St. John's, and Harbour Grace. Abbé Jean
Baudoin, who travelled with d'Iberville as his chaplain during the
raids on Carbonear and Harbour Grace, Newfoundland, said this
about the difficulties the weather presented:

January 17th, 1697

We joined them [M. d'Iberville and his men] making in one day the distance which they had scarcely made in two, the woods being covered in snow so that it was nearly impossible to get out of it. Canada has nothing like it.

January 18th, 1697

The trails are so bad that we could find only twelve men to break trail. Our snowshoes are burst with the frost and icy rocks and in the snowy woods one makes false steps. One cannot with all this keep from laughing to see some falling here, others there, into the snow. Montigny [Jacques Testard de Montigny, officer in the colonial regular troops in Montreal] fell into a river and left his musket and sword in order to save his life.[30]

1776

In 1776, Church of England clergyman and Methodist missionary Reverend Laurence Coughlan inserted in *An Account of the Word of God in Newfoundland, North America* this description of winter conditions:

The Winters in Newfoundland are very severe, there being great Falls of Snow, and hard Frost; the Houses there are mostly very disagreeable to those who are not used to them; in general; they are all Wood; the Walls, so called, are Studs put into the Ground close together, and between each, they stop Moss, as they call it, to keep out the Snow; this they cover with Bark of Trees, and put great Clods over that; some are covered with Boards: In such Houses I have been, and in the Morning my Bedside has had a beautiful white Covering of Snow; my Shoes have been so hard frozen, that I could not well put them on, till brought to the Fire: But under all this, I was supported, seeing a glorious Work going on.[31]

1790

The thermometer in Hopedale on January 6 showed -40°C, according to the Moravian mission's weather data.[32]

The Moravian Brethren, formed originally in Bohemia under reformer Jan Hus, were driven almost to extinction during the 30 years of war in central Europe (1618–48). They had a renaissance of sorts when they migrated to Saxony, Germany, where Count Nicolaus Ludwig Von Zinzendorf gave them refuge. From there they branched out all over the world to do missionary work—and record weather data. In the 1700s they travelled to Labrador and brought their Christian beliefs to those they called "Eskimos." They subsequently established missions at Nain, Hoffenthal (Hopedale), Okkak (Okak), Ramah, Zoar, Hebron, and Makkovik. Men with scientific minds, the Moravians recorded their own observations on the weather, climate, air pressure, temperature, and wind direction in the small coastal communities, and they collected meteorological data from the eighteenth to the twentieth centuries.[33] As of 2014, Moravian congregations still exist at

Hopedale, Nain, Happy Valley, North West River, and Makkovik. The missions have all gone; those at Okak, Ramah, Zoar, and Hebron closed in the late nineteenth and mid-twentieth century.

1796

In 1796, Methodist preacher William Thoresby, who brought the Word of God to the people of Newfoundland, described in his diary the weather and the disasters it caused, and how he coped.

> Jan. 23rd, 1796
> This day [in Harbour Grace] has been the most stormy for wind and snow I ever saw; and as it was so exceeding stormy, I expected none at the evening preaching; but to my great surprise, about forty persons came through the snow up to the waist: I hope they did not come in vain. The two last places I have had to sleep at, the snow blew into the rooms and even on my face, I have thereby caught a severe cold....

> Jan. 26th, 1796
> ... I parted with my friends in this harbor in peace and seven men rowed me ten miles in a skiff to Harbourgrace; they had to beat through much ice and the frost was very severe. I lay with seven great coats around me at the bottom of the boat and it was with difficulty that I escaped being burnt with the frost.[34]

1859

Roman Catholic bishop John Thomas Mullock of St. John's (1850–68) recorded these weather statistics for 1859:

Highest temperature, July 3: 32.2°C

Lowest temperature, March 3: -13°C

Mean maximum barometer pressure: 29.74 inHg

Rain for the year: 162.4 centimetres

Maximum quantity in 24 hours: 5.3 centimetres

Wind NNW and WNW: 200 days

Wind NE: 25 days

Wind W and WSW: 38 days

Wind SSW and SE: 102 days

Rain: 110 days

Snow: 54 days

Thunder and lightning: 5 days

Mullock concluded that the local weather was invigorating and salubrious with a lack of indigenous diseases. But for those who would criticize the Newfoundland climate, he offered this:

What an awful climate, they will say, you have in
Newfoundland; how can you live there without the sun, in
a continual fog? Have you been there, you ask them? No!
They say; but we have crossed the Banks of Newfoundland.
How surprised they are then when you tell them that for
ten months, at least, in the year, all the fog and damp of the
banks goes over to their side and descends in rain there with
the south-westerly winds, while we never have the benefit of
it unless when what we call the "out" winds blow. In fact the
geography of America is very little known even by intelligent
writers at home, and the mistakes made in our leading

periodicals are frequently very amusing. I received a letter from a good intelligent friend of mine some time since in which he spoke of the hyperborean region of Newfoundland. In my reply, I dated my letter from St. John's N. lat. 47 [degrees] ...[35]

By comparision, London, England, is at 51 degrees latitude.

1892

Occasionally, the thermometer can surprise us. On January 16, 1892, a temperature of 15°C was recorded at St. John's. One hundred and four years later, in 2006, the record for that date was broken when the temperature at St. John's rose to 15.7°C.[36]

1896

St. John's newspapers reported that January 1896 started out as mild with football matches and flowers sprouting in the gardens:

> Some gentlemen were curious enough to examine the flowers in their gardens yesterday and found that so far from having been injured in the recent frost and snow, they were actually sprouting. This reminds us that Captain Costigan and other Labrador men always plant their cabbage seed in the fall of the year, and when they go down in the spring, the young plants are all ready for transplanting.[37]

Milkmen were having difficulty getting around with horse and sleigh due to the bare spots in the roads; the route between

CHILDREN SLEDDING, LOCATION UNKNOWN

Rev. H.M. Dawe Collection, United Church Conference Archives, St. John's

Torbay and Pouch Cove was entirely bare and unfit for runners. One reporter noted that individuals fond of skating and coasting would love to have "old Jack in a 'biting' mood."[38]

By January 22, however, the snow had arrived in St. John's and was proving dangerous to pedestrians or, at the very least, embarrassing. As that day's *Daily News* reported:

> Yesterday afternoon a man was coming up Theatre Hill when a shout came from a bob-slide full of boys, "Out of the road Mister!" Almost at the same moment he was saluted in the rear with the cry of "look out there!" and turning to face the

new danger he found that a horse and sleigh was almost upon him ... before he had time to do anything, the bob-slide had given him a gentle hint that he didn't own the street; his legs appeared to forget that the perpendicular was their proper position, and he was embracing the two foremost boys much against his will.[39]

The next day, January 23, began a four-day stormy period on the east coast of the island:

The Storm raged all last night ... Though one would scarcely think it, there were numbers of the fair sex out plunging and tumbling through the snow drifts, but whether on pleasure or business bent, it was hard to say. One thing is certain, if they ventured out to test the value of the work done by the latest thing in "curling irons" they were sadly disappointed, for no frizz could stand the blizzard ... In some streets this morning the inhabitants got up, opened their front doors and looking out saw—well they didn't see very far, because the snow was above their doors, while the people on the other side hadn't the semblance of a drift ... several gentlemen living in the suburbs had to come to the city on snow shoes to-day, the drifts being twenty feet deep in some places. A prominent business man of the city ... found such a bank of snow in front that it would take a week to get through it and having tried the back discovered that it was nearly as bad. There was no help for it. A tunnel must be cut and this he proceeded to do. His man commenced at

MAN IN SNOW TUNNEL, ST. JOHN'S, 1918

City of St. John's Archives

the outside while he worked from within and now he has a charming covered way extending about 50 feet from the back door.[40]

But this *Daily News* writer did find something positive to report about the wintry weather:

There were some pretty views to be seen around the city and suburbs yesterday, but they were nothing to the effect produced by today's bright sunshine. It was a truly magnificent sight to stand on one of the west end wharves just after the sun had risen this morning and gaze over the harbor.

A SNOWBOUND TRAIN, CA. 1900

The Rooms Provincial Archives Division, VA 19-213

> Every rope on every vessel in port was a line of dazzling silver,
> and even the spars and hull appeared to be sheathed in the
> precious metal.[41]

Train travel across Newfoundland was difficult when huge walls
of snow blocked the rail lines and it was also dangerous work
getting them cleared. One railway worker, Robert von Stein,
had a narrow escape: as he jumped from a snowbank onto the
engine, he slipped, and slid down under the wheels. A.S. Noble
promptly jumped to the rescue and pulled him safely on board.
Stein said he "wouldn't mind having his legs cut off if the engine
only left him his head." In heavy snow, the plows on the front of
the trains had to butt the snowbanks to make a way through. In
one case, No. 10 with its "Eagle wing" snowplow started from St.
John's, with No. 7 following close behind. They met the snow near

St. Anne's, about 7 or 8 miles out of St. John's. No. 10 accelerated and drove the Eagle against an enormous snowbank at a speed of about 64 kilometres per hour, but could not get through; it had to be pulled out by No. 7. After an hour of continuous butting, No. 10 managed to cut a passage. Some cuts made by train and plow were 457 metres long and almost 8 metres deep.[42]

1942

On January 29, a correspondent for the *Evening Telegram* wrote this alliterative description of the weather St. John's had just experienced: "Following a short but snappy snow storm yesterday evening, rain came on in torrents last night and this morning many city streets were miniature lakes of knee deep slush."[43] Many St. John's neighbourhoods were affected by this storm, particularly the west end in the vicinity of Victoria Park and the south side of St. John's harbour. Gushing rivers washed dirt and debris down the hillside to form large muddy lakes at the bottom of Leslie and Springdale streets. Shops and houses on the south side of Water Street flooded and, as the Waterford River overflowed its banks, the St. John's Gas Company plant at the bottom of Patrick Street collected water to a depth of almost 1 metre. The water put out the gas fires in the plant and temporarily shut it down. With the gasworks flooded and the Petty Harbour power plant also out of commission, Newfoundland Light and Power Company asked the city's residents, who still had power, to conserve electricity. Several hours later the utility company managed to hook up another power plant at Pierre's Brook.

A VIEW OF THE GASWORKS, WITH ST. PATRICK'S CHURCH ON THE RIGHT, IN THE WEST END OF ST. JOHN'S, CA. 1914
ASCD

During the power outage, city streetcars ground to a halt and near Victoria Park debris buried the tracks for a distance of more than 100 metres.[44] Most of the houses that lined the road opposite the park lay in the path of the flood. Many homeowners had water over the first floor of their houses, and John Coady and his family, who lived in a basement apartment at 255 Water Street, had to get out quickly when their furniture started floating. On Alexander Street, silt from the flood buried a fire hydrant.[45]

LONG BRIDGE, CA. 1915

City of St. John's Archives

Long Bridge, which ran across the harbour from behind the railway station to the Southside Hills, became jammed with ice and timbers that had broken loose from a boom of wharf piles. The bridge held until the ice and timber passed through. Water rushing down the Southside Hills brought gravel and rock to the roadway. To add to the disaster, a sewer pipe in the area burst, flooding Southside Road to a depth of almost 0.5 metres.[46]

Farther east in the capital city, Long Pond Bridge gave way.

Water levels at Long Pond rose to their highest in 35 years, flowing over the bridge at the outlet and making 30 metres of the road on each side impassable without hip waders. The water in Leo Parrell's ice house at the side of Long Pond was waist-high, and he lost several hundred blocks of ice, which were swept down the river by the rushing waters. The loss was catastrophic—Parrell had cut and carted the ice all the way from Kent's Pond, about 1.5 kilometres away. With the high water pressure, Long Pond Bridge finally gave way, although enough of the bridge remained that farmers from Nagles Hill who had come into town could still safely lead their horses over it.[47]

As Rennie's Mill River spread out across the fields, trees and fences disappeared under 1- to 1.5-metre-high water. Feildian Grounds, which lay close to the riverside, turned into a lake with water 1 metre or so deep almost all the way to Quidi Vidi, while the roadway around the foot of the lake was impassable due to the depth of the water. On the upper and lower sides of Quidi Vidi Lake basements flooded and 1 metre of water covered the floor of the Star Boat House,[48] one of several boathouses on the lake.

Council workers were out in droves the next day, cleaning up dirt and debris from the rain-ravaged streets. The flood waters left a pile of cans, bottles, and other rubbish by the fence at Victoria Park.[49]

Making the situation worse, temperatures dropped suddenly and streets became sheets of ice. Water covered the ice in many places, and residents soon discovered a few deep pools here and there, increasing their risk of injury. At the time—the height of World War II—St. John's was in the middle of a test blackout. At night, the city lay in darkness. With no lights, residents found it difficult, and too dangerous, to move around after nightfall.[50]

The *Evening Telegram* editorial the day after the storm focused on conserving water and the possibility of water shortages due to the heavy usage put on the supply by visiting American and Canadian forces.[51]

The rainstorm mainly affected the southeastern sections of the island. North of St. John's, sleet damaged telegraph poles and lines. Ice, up to 2.5 centimetres thick, coated the wires and, between Heart's Content and Carbonear, many poles came down. West of Bishop's Falls, in central Newfoundland, a raging snowstorm necessitated push plows to clear railway lines on the Gaff Topsails, a plateau of high land approaching the west coast. Weather conditions disrupted train service across the island.

WEATHER SCIENCE

Many winter storms approach Newfoundland from the southwest with winds that blow counter-clockwise around the low pressure system. As most precipitation occurs ahead of a storm centre, when the low is to the south of the island the resulting easterly winds and relatively mild temperatures bring rain, freezing rain, or snow. As the low moves to the north, because of the

RING AROUND THE MOON IN PORTUGAL COVE-ST. PHILIP'S
Photo by Sue Willis

counter-clockwise movement, the direction changes to a westerly wind, with colder, drier air. As a result, the temperature usually drops and precipitation tapers off. Sometimes, however, strong westerlies blowing across open water bring heavy snowfall and blizzard conditions.[52]

TRADITIONAL WEATHER LORE

A ring around the moon is a sign of rain or snow on the way.

A halo or corona around the sun or moon occurs in the presence of clouds that are composed of either ice crystals, water droplets, or a mixture of both. The first clouds to appear in the sky in advance of a storm are high wisps of cirrus clouds,

ST. JOHN'S SNOWSHOE CLUB, GOVERNMENT HOUSE GROUNDS, CA. 1890

ASCD

called mares' tails, composed of ice crystals. As the storm advances, the cirrus clouds thicken and spread across the sky. A ring around the sun or moon may appear. When light passes through ice crystals in the clouds, it may be refracted or bent, and a halo forms. Sometimes a rainbow effect occurs if the light is refracted through drops of water in lower lying clouds. The more ice crystals, the brighter the halo, and therefore the closer the storm.[53]

SO HOW DOES ONE DESCRIBE THE
NEWFOUNDLAND WINTER?

David Phillips summed up Newfoundland winter with this composite of Newfoundland weather terms:

> Woolly-whipper warning ends in early marnin' however citizens should be wary of the slob on the tickle. Today, mawzy with weatherish periods, the inwind will back to an outwind and expect isolated scuddies early this afternoon ... Tonight ... it will turn duckish with a screecher from the north and a batch of snow ... Wednesday ... becoming more civil, stun breeze by mid day ... Be alert to falling conkerbills, bubbles, and bally-cutters.[54]

1965

On January 1 a five-day snowstorm buried the Cartwright area of Labrador.[55] Snow was sparse on December 31, New Year's Eve; only 3 centimetres lay on the ground. That changed when the storm hit. Snowfall amounts for the next five days: 15.2, 56.9, 58.9, 29.2, and 21.8 centimetres. On January 6, another 3 centimetres fell. A total of 182 centimetres—almost 2 metres—of snow fell. High winds in the 60–70 kilometres per hour range churning up the heavy snowfall resulted in a miserable few days for residents.[56]

In 2010, Environment Canada's annual "Weather Winners" ranking declared Corner Brook the snowiest city in Canada, with an average annual snowfall of 422 centimetres. The snowfall in

WOMAN IN WINTER DRESS, CA. 1920

City of St. John's Archives

some of the smaller centres surpassed that, with St. Anthony at
505, Churchill Falls 465, Happy Valley-Goose Bay 459, Gander 443,
and Deer Lake 425 centimetres.[57]

1966

"Avalon digs out after paralyzing storm," the *Evening Telegram*
headlined its January 11 paper. Purported to be the worst storm
in 30 years, this storm smothered the Avalon Peninsula with 45.7
centimetres of snow on January 9 and 10. Drifts of 3 to 5 metres
formed in hurricane winds of 138 kilometres per hour, knocking
out power and causing widespread damage. As the blizzard
raged, lightning flashed across the sky, but residents did not hear
thunder. Perhaps the roaring of the wind drowned it out. In the
midst of the chaos, the power went out, and a large part of St.
John's remained in darkness for 12 hours—difficult conditions
for most people, but the two-day school closure likely delighted
schoolchildren. At the height of the storm, our Lady of Lourdes
School on Nagles Hill, at the northern outskirts of St. John's,
burned down, under suspicious circumstances. As high winds
whipped the fire into an inferno, firemen were hampered by the
winds, the snow, and a lack of water, able to draw only a small
amount from a nearby stream. Within two hours, the fire had
left a burned-out shell of a building and $500,000 in damages.

In the Outer Battery, at the mouth of St. John's harbour, several
nervous families vacated their homes, worried that an avalanche
of snow might thunder down the steep cliffs above them, as it
had twice before, in 1921 and 1959.

VINCE AND ANDREW MELVIN CROSSING THE CRIBBAGE BRIDGE AT LA MANCHE, CA. 1952

Maritime History Archive, Memorial University of Newfoundland (MHA)

The storm also cancelled flights in and out of St. John's and delayed trains for 24 hours. In some coastal communities, heavy seas washed out those train tracks that ran close to the shore. Buses were back on city streets by Tuesday, January 11, but getting around was still difficult because many streets had only one lane. Joseph Smallwood, premier of Newfoundland, finally made it into St. John's on Tuesday after being snowbound for three days at his home on Roche's Line about 80 kilometres away.[58]

The effects of this storm were far-reaching. On the Burin Peninsula, a bridge and section of the causeway at St. Lawrence

washed out, and a newly constructed breakwater collapsed. At
Marystown, the tides rose higher than usual and cracked the new
government wharf as well as damaged other private wharves.
The Avalon and Burin peninsula highways, and the Trans-Canada
Highway as far west as Gander, were blocked with snow. Corner
Brook, which escaped the brunt of the storm, received only 3.8
centimetres.[59] Even though the west coast was not as severely hit
as the east coast, the *William Carson*, the ferry on the Port aux
Basques to North Sydney run, was tied up for two days.

In 1966, La Manche, appropriately named for its long, narrow,
high-sided shape, meaning "sleeve" in French, was a small fishing
village located 53 kilometres south of St. John's. Settled by several
English families in 1840, La Manche became known as one of the
best fishing coves on the southern shore. In the early years of
settlement, the French used La Manche as a hiding place after
their raids on Ferryland and St. John's. During the 1950s and
1960s, the Newfoundland government, under Premier Smallwood,
paid people living in small isolated communities to relocate to
larger settlements. La Manche residents came under government
pressure to move, but they refused. On January 25, 1966, however,
their fate was decided by a severe winter storm that hit the
eastern part of the island.[60] High winds with estimated speeds
of 121 kilometres per hour[61] caused an enormous storm surge,
which rushed into the narrow rock channel, dismantling houses
and washing away flakes, boats, stores, a recently completed
boardwalk around the harbour, and a suspension bridge that
joined both sides of the harbour.

Phillip Melvin, long-time resident of La Manche, lost his two-storey, 15-by-6-metre stage and a 6-by-6-metre splitting stage, all built of thick planks—the sea smashed the structures and washed them away. His house shifted off its foundations and only barely escaped being washed away with the stages. Of the nearby suspension bridge, only the steel bridge cables remained.[62] After the storm laid waste to La Manche, the residents opted for resettlement rather than rebuild their shattered properties.

A few days after the disaster at La Manche, another storm, on January 28, crashed into the Avalon Peninsula with 11-metre waves and howling 121-kilometre-per-hour winds.[63] The wind and rain caused damages severe enough to financially cripple several St. John's fishermen with losses amounting to nearly $200,000. Mr. Riche in the Battery lost his 24-metre stage, a boat, three engines, a herring seine, four salmon nets, and a quantity of lumber. Mr. Wareham, also of the Battery, lost a stage that had been built in 1913.[64] The storm uprooted trees, smashed windows, tore roofs off houses in St. John's, and blew cars off the Trans-Canada Highway. Losses were also incurred on the south side of St. John's harbour, in Petty Harbour, Quidi Vidi, and Conception Bay communities, and along the Southern Shore. In Petty Harbour, 40 fishing stages collapsed into the sea and fishermen lost almost all of their shoreside properties. Mr. Whitten of Petty Harbour estimated the damages to be close to $1 million.[65] In Brigus, one fisherman lost his boat, all his gear, and other property, valued at $15,000.[66]

WEATHER SCIENCE

Winds blow in a clockwise direction around high pressure systems. These systems are typically associated with clear and dry weather, light winds, and descending air. An interesting fact about the descending air of a high pressure system is that it tends to inhibit cloud formation, which in cool weather can act as an insulator. In a low pressure system, the winds move counter-clockwise, and the air rises. As it does, it cools, and the moisture in the air condenses to form clouds. When the clouds are heavy enough, moisture falls to the ground in the form of rain or snow. Low pressure systems are associated with windy, wet, and cloudy conditions.[67]

1977

The lowest sea level pressure ever recorded in Canada occurred at St. Anthony on the Northern Peninsula of Newfoundland on January 20, 1977, when the barometric pressure dropped to 94.02 kilopascals[68] (average sea level pressure is about 101.32 kilopascals). This pressure drop was the prelude to one of the most powerful storms of the winter, pounding the island with winds up to 161 kilometres per hour. In St. Anthony, 31 centimetres of snow fell. By the following day, 83 centimetres of accumulated snow lay on the ground.[69]

The 10 centimetres of snow that fell in St. John's was a minor event when compared to what fell on the rest of the island. Deer Lake recorded 45 centimetres, and 25 to 29 centimetres fell in central Newfoundland. Cancelled flights left people stranded at airports, and power lines across the island were disrupted, forcing the Newfoundland Light and Power Company personnel to work around the clock to catch up. Schools and businesses closed, and the ice-cold temperatures made it miserable for anyone setting foot outdoors.

In 1977, the temperature was not measured in Wind Chill Equivalent Temperatures but in the rate of cooling of exposed flesh in watts per square metre. In St. John's that day, the wind chill factor was 1,700 units, the level at which exposed flesh will freeze after prolonged exposure. The rest of the island was only slightly warmer, with readings of 1,600 and 1,500 units.[70]

In the midst of this biting cold, a fire in the Southside Road substation disrupted the city's entire electrical system. It was not fully restored until 6:30 p.m. the following day.[71] Newfoundland Power linesman Frank Cole, from Signal Hill, along with many others, worked a difficult and miserable 24 hours fixing lines that had been damaged by the storm. The exhausted Cole wore three sweaters, two coats, insulated boots, and rubber trousers.

During the outages, St. John's residents visited friends and neighbours to stay warm. Those with oil stoves weathered the storm well; those who had recently switched to electric heating rued the day they had. They probably went to bed that Thursday night with their clothes on and blankets piled high. One family, who went visiting to stay warm, had a baby recovering from heart surgery. Unaware of the chaos of the storm, the baby spent a comfortable night in a bureau drawer at the host's house. Those small shops that managed to stay open did a roaring trade as people poured in to purchase necessities. As some pizza parlours and takeouts were able to keep cooking, people could get hot food after the power went out.[72]

Despite the high winds, the Battery escaped this storm almost unscathed, with only about $5,000 damage to boats, gear, and two wharves. Mavis Wells on Outer Battery Road said that she and her neighbours had no problems and that there were no disruptions in power or telephone services. "It was all right in this area," she stated.[73] Other areas that lost power for short periods of time included Pouch Cove, Carbonear to Holyrood, Bay de Verde, Coley's Point to Port de Grave, and the Cape Shore on the southern Avalon Peninsula.

The most severe repercussions of the storm were felt in Badger, a town in central Newfoundland situated in excellent timberland, where the Exploits River, Little Red Indian Brook, and the Badger Brook meet. The geography of the area made it prone to flooding and in this case the banks of the Exploits River, steadily rising since January 17, were taxed to the limit with an accumulation of 30 to 35 centimetres of snow. An ice jam in the river finally caused the waters to break through, flooding Badger and making it necessary to declare a state of emergency. The flooded Exploits left as much as 2 metres of water in some parts of the town. Thirty-eight families were evacuated and either stayed with friends or neighbours or relocated to Grand Falls. Badger residents had tried to prepare for the impending disaster by sandbagging some areas, but on the night of January 20 their efforts could not stop the swollen river from bursting its banks. The water finally found a channel out, but contamination remained a concern, as sewers had backed up. A state of emergency remained in effect until the morning of January 24, restricting cars, public gatherings, church services, and other activities.[74]

In some areas of Badger, the only means of transportation was by canoe. Mabel Day referred to the flood as a "shocking mess." The Day family managed to keep most of the water out of their home for a while by running three pumps, but eventually they were forced to move out. "You'd open the door and the water would come in like a brook," Day said. Earle McCurdy, who later became the president of the Fisheries, Food, and Allied Workers Union, was, in 1977, a reporter for the *Evening Telegram*. He described how the town had considered blasting the ice jam with dynamite but, fearing that that might aggravate the situation, decided against it. Demolition experts from the Armed Forces in Gagetown, New Brunswick, came to Badger to give advice in the event of a worsening situation. A Grand Falls radio station launched an appeal for food, blankets, clothing, and money, and the town received donations from as far away as Toronto. By Monday the station had collected over $16,000.[75]

Similar flood scenes played out far from Badger in Placentia. In some parts of Placentia the water rose to 2 metres when the river overflowed, washing against front porches, filling basements, and covering kitchen and living room floors.

TRADITIONAL WEATHER LORE

If there is enough blue sky to make a pair of sailor's pants, the weather will turn fine.

My elderly aunt often quoted this adage on cloudy, foggy days when a small patch of blue sky broke through the grey. We children were confused, however, as she never specified the size of the sailor—we wondered how much blue sky would be needed.

A CLOUDY SKY AT ST. GEORGE'S LAKE, SPRUCE BROOK, ON THE WEST COAST OF NEWFOUNDLAND, 1935

The Rooms Provincial Archives Division, VA 8-19 / Alfred Cooper Shelton

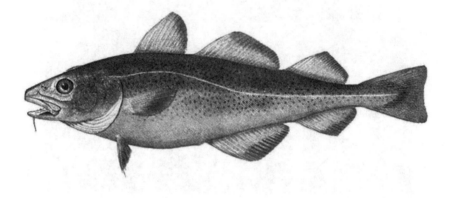

GADUS MORHUA—COD

US National Oceanic and Atmospheric Administration (NOAA)

FEBRUARY

TRADITIONAL WEATHER LORE

A lot of snow and ice and a long winter mean a good fishing season to come.

This phrase may contain some truth. A cold winter lowers the water temperature, and significant snow and ice raise the water levels in rivers and streams, which allow plankton to float out to sea and feed the cod. With a regular food supply comes a strong population of cod.[76] Tiny cod larvae feed on plankton for several months after they have used up the food supply in their yolk sacs. Young cod fry feed mainly on copepods, amphipods, and other small crustaceans in the plankton.[77]

1583

Stephen Parmenius wrote to Richard Hakluyt the Younger from Newfoundland:

> ... but how cold it is in the wynter, the great heapes, and mountaines of yce, in the middest of the Sea have taught us: some of our company report, that in may they were sometimes kept in, with such huge yce, for 16 whole days together ...[78]

1613

Henry Crout, in Cupers Cove, kept a daily weather diary from

A GOOD FISHING SEASON ON THE BANKS, DATE UNKNOWN

NOAA

September 1, 1612, to May 13, 1613, recording the weather morning, noon, and night. On February 10, 1613, he recorded this:

> In the morning the wind at west not much wind, the sune shininge all the daie verie pleasaunte and warme thought [though] a litell cold but all our men did work in the woodes with great Content but in the after noone abowt 4 of the clock the skie verie cleare begynninge agayne to Freesse very hard[.]

After dynner died one of our Companie called Edward
Hartland a taillor and prentice for the companie.

This morning the litle pound betwine the house and the
Brue house and allso the harbour was all taken over with a
crayme of Ice this night the wind at west north west litle
wind but Freessing hard all the night.[79]

Six days later, Crout reported more moderate weather: "Sune
shininge verie warme untill noone which desolved the snowe
and ice verie much very moderate weather as hart cold [could]
wishe which putteth everyone in hope that winter is now paste
for this year ..." He spoke too soon. The snow continued to fall
until May 13.[80]

1629

Colonists at Ferryland, in the Colony of Avalon, experienced
a hard winter in 1629. Lord Baltimore probably shivered as he
wrote these words to his friend Sir Francis Cottington: "in this
woeful country, where with one intolerable winter, we were almost
undone. It is not to be expressed with my pen what wee have
endured." Baltimore described their hardships in a letter to King
Charles I of England:

From the middest [middle] of October, to the middest of May
there is a sadd face of wynter upon all this land, both sea and
land so frozen for the greatest part of the tyme [time] as they

are not penetrable, no plant or vegetable thing appearing out of the earth untill it be about the beginning of May nor fish in the sea besides the ayre [air] so intolerable cold as it is hardly to be endured ... my howse [house] hath beene an hospital all this winter; of 100 persons, 50 sick at a time, myself being one and nyne or ten of them dyed.[81]

Many Ferryland colonists suffering from scurvy in the winter of 1629 died before they discovered that eating raw turnips would have cured this vitamin C deficiency.

TRADITIONAL WEATHER LORE

February 2, known as Candlemas Day or, more recently, Groundhog Day, is an important day for forecasting the coming months, as this rhyme suggests:

> *If Candlemas Day be fair and fine, the worst of the winter is*
> *left behind*
> *But if Candlemas Day be dull and glum or grum, then the*
> *worst of the winter is yet to come.*

This version of the rhyme is different than that in the rest of North America in which, if the groundhog sees its shadow— if the sun is shining—the worst of winter is yet to come. In medieval England, Candlemas Day marked the halfway point of winter. If only half the grain and other supplies had been used by then, there would probably be enough to last the rest of the winter.

Two other versions of this rhyme exist in Newfoundland. From Seal Cove, Fortune Bay:

> *If Candlemas Day is fair and fine, half the winter is left behind.*
>   *If the day is dark and grum, half the winter has to come.*

A less popular version came from Riverhead / Harbour Grace:

> *If Candlemas Day is fine and fair, there'll be two winters in*
>   *one year.*
> *If Candlemas Day is dirty and rough, the rest of the winter*
>   *will be fine enough.*[82]

Which is true? The beauty of having two opposite outlooks is that you can be proven right no matter what the weather on February 2.

1796

Thoresby summed up the Newfoundland climate as intolerably hot in summer and intensely cold in winter:

> February 22nd, 1796—The last night was the most sharp and severe for frost that they have had in Newfoundland for ten years back; the Bay was froze over in a very short time; and my bed was covered with hoary frost.

SKATING ON THE WATERFORD RIVER NEAR THE RAILWAY STATION, 1920—21

City of St. John's Archives

February 23rd—Though the morning was very severe, the wind exceeding high, and the snow flaggs [flakes] flying in every direction, I set off in the storm and travelled ten miles almost up to the knees in snow, to Clor's Cove. I changed my linen, (which was as wet as if it had been drawn through water) and after refreshing myself proceeded to Witson's Bay; none can tell what I suffered in this journey, but God, and myself: through the goodness of my God, I got safe to Mr. Perey's, where I preached and slept. I am heartily glad that God has begun to work in the heart of Mr. Perey's son.

Thoresby discovered that preaching to the errant souls of early Newfoundlanders came with its challenges. He wrote on February 26, 1796:

> The last night the Bay was froze over in the space of an hour; and across that part of the Bay where I am at the present is not less than six leagues over, how astonishing ... In the forenoon I read prayers and preached in the church, and though I had two pair of worsted gloves on my hands, two pair of stockings and a pair of buskins [knee or calf length boots] on my legs, it was with difficulty that I escaped being burnt with the frost.[83]

1797

In Labrador, the cold became so severe in January 1797 that in Okak the thermometer registered -37.8°C and remained there for the whole of February.[84]

TRADITIONAL WEATHER LORE

When the snow crunches underfoot, the next day will be mild.[85]

There might not be a scientific explanation for this one. Crunchy snow indicates that there has been some thawing and refreezing, which gives a crust of snow. Perhaps it is a sign of spring on the way.[86]

Newfoundland historian and author Paul O'Neill (1928–2013) recalled boyhood winters in St. John's in the 1930s and 1940s:

> We'd go sliding down Fraser Lane, that small lane that cuts down from Circular Road to Empire Avenue. We could slide down the lane, across Empire, and down Rennie's Mill Road. In the winter, no traffic was allowed on Empire Avenue, so the kids could slide safely.
>
> We went skiing at North Bank Lodge, which at the time was next to the present day Fluvarium on Nagles Road. We skied down the hill to Long Pond.
>
> We went skating too, on Quidi Vidi Lake. My mother would always say "Be careful when you go down to the pond. It's not always frozen over." A swimming area in summer, when it froze in winter it was cleared by the city, and became a skating area. It was located off boathouse lane, the small road that runs past the present boathouse.[87]

1921

On February 8, 1921, an avalanche crashed down Signal Hill onto the small village of the Battery. As fishing flakes and fishermen's sheds near the water's edge collapsed under the weight of the snow, even more dramatic events unfolded farther up the hill. The avalanche moved Alfred Wells's house 3 metres off its foundation and pushed the roof down into the bedroom, injuring Alfred and

A YOUNG GIRL ON THE STEPS OF HER HOUSE ON MUNDY POND ROAD, CA. 1921

Private collection, S. Roberts

Annie Wells. Although he suffered broken ribs, Alfred managed to get himself out and free his two-year-old son, Sam, from his crushed crib. He then freed his pregnant wife, Annie, who had sustained a back fracture, and infant daughter, Geneva, who was nearly smothered under the snow. The family crawled through a hole into the night blackness. Dressed only in nightclothes, and with the storm raging around them, they made a horrendous journey down the hill to Alfred's brother's house. Annie was taken to hospital, where she stayed until the birth of her daughter Elizabeth several months later. Homes belonging to the Morris family, the Edgecombes, and Moses Piercey were also damaged, but there was no loss of life.[88]

Ten days later, a second snowslide in the Battery killed Albert Delahunty. Searchers found his body 70 metres below his house on Signal Hill, still holding his dinner pail. The exact location of Delahunty's house is not known, but it was probably located close to the Queen's Battery, where no houses remain. Delahunty may have inadvertently triggered the slide by leaving the path and breaking through a cornice as he made his way to work in the raging snowstorm.[89]

In St. John's, the heavy snowfall piled drifts 2 to 3 metres high. Because the city had no mechanical equipment to remove the snow, residents depended on shovellers and teams of horses with box carts.

ST. JOHN'S SNOW SHOVELLERS, DUCKWORTH STREET, LOOKING EAST, DECEMBER 1904

City of St. John's Archives

WEATHER SCIENCE

Environment Canada defines *wind chill* as the cooling sensation caused by the combined effects of temperature and wind. In cold weather on a wind-free day, body heat creates a "boundary layer," a thin film of air between skin and clothing that acts as an insulator. When the wind blows this heat layer away, our bodies use up energy trying to replace it. As the heat layer continually gets blown away, our skin temperature drops, and we feel cold.

The wind will evaporate any moisture on our skin, cooling it even further, and can lead to frostbite and hypothermia.

Since 2001, wind chill has been measured in Wind Chill Equivalent Temperature. According to the standards set that year, a wind speed of 60 kilometres per hour, which would cause trees to bend, and be difficult to walk against, combined with a temperature of -10°C would feel like -23°C. A wind speed of 40 kilometres per hour, which would cause small trees to sway and large flags to extend and flap strongly, along with a -10°C temperature would feel like -21°C.[90]

1942

> Warship and Freighter Lost near St. Lawrence.
> Heavy loss of life owing to raging gale.
> People of St. Lawrence praised for rescue work.
> —*Evening Telegram*, February 24, 1942[91]

At the height of World War II, three US naval ships left Maine, bound for the American military base at Argentia, Newfoundland, on the south coast. The destroyer USS *Truxtun* and command ship USS *Wilkes* were escorting the USS *Pollux* as she transported cargo for the war effort through the dangerous waters of the North Atlantic. As the vessels neared Newfoundland, the wild and destructive force of a violent winter storm, rather than enemy subs, attacked them. A navigation mistake brought the ships in the convoy too close to the land,

and conditions became treacherous as they approached the jagged coastline.

From the time the ships left New England, those on board had not seen the sun or the sky for about three days due to inclement weather—this was a problem because navigation was done by dead reckoning (a technique that uses the stars to set location). As the ships approached Nova Scotia on the morning of February 17, there was a slight break in the weather, and Lieutenant Arthur Barrett, the navigator on the *Wilkes*, made an error when he finally got a chance to set their position. He calculated the position of the star Antares as west instead of east of the meridian.[92] As the ships approached Newfoundland, the course deviation worsened and it was not until they neared the Banks and took a sounding that Lieutenant William Grindley, the navigator on the *Pollux*, realized something was wrong. When he asked his commander to make changes in the course for Argentia, his superior, not convinced any change was necessary, agreed to a slight, 10 degree course change, but it was not enough.

By noon, the weather had deteriorated, with the barometer at 29.88 inHg, and falling.[93] By 11:30 p.m., it was blowing a gale.

The ships usually communicated to each other with signal lights, but they could not see each other through the blinding snow and sleet and they had no way of warning the others of any dangers ahead. As freezing rain and sleet covered the ships, it disabled their equipment: radar became unreliable and radio contact was rendered useless.

By 2 a.m. on February 18 the wind had changed direction from southeast to east. With zero visibility, the sailors on watch had no idea how close they were to the coast until they saw the cliffs almost on top of them. One by one, the ships foundered, grounding on the jagged rocks. Within 10 minutes, all three ships ran aground off Newfoundland's Burin Peninsula. The *Wilkes* grounded at 4:09 a.m. on the southwest corner of Lawn Head, managed to break free, and limp away, but sleet and high winds hammered the remaining vessels. The *Truxtun* went aground at 4:10 a.m. on February 18, wedged between the rocks just off Chambers Cove, 230 metres from a small beach at the base of perpendicular cliffs. Violent waves and freezing spray hampered efforts to get rafts or towlines set up, but after several men lost their lives, two brave souls finally made it to shore with a line. Using that line, 24 men made it off the ship before they lost their raft. Two of the men from the *Truxtun* climbed the 100 metres of wet, icy cliff only to find a barren snow-blown wasteland at the top—not a human in sight, only an abandoned shed. One man took shelter in the shed, too exhausted to go any farther; the other managed to get to the Iron Springs Mine about 2 kilometres away, just outside St. Lawrence. By 8 a.m., the barometer had dropped to 28.85 inHg[94] and the wind had shifted to the south.

Gus Etchegary, president of Fishery Products Ltd. from 1976 to 1984, was born and raised in St. Lawrence. Gus, along with his

THE RESCUE OPERATION AT CHAMBERS COVE
ASCD

brother Theo, and their father, Louis, a mill superintendent and
mine captain, took part in the rescue effort. Louis Etchegary
happened to be standing outside in a huddle with several other
mining supervisors at 6:30 a.m. planning the day's operations
when, through the snow, they spotted a man running toward them.
Amazed, they heard his tale, and the miners, along with people
from the community, rushed to the scene of the catastrophe. They
saw men in the water clinging to debris and others still holding on
to the ship.[95] As the rescue line had broken, the sailors had to jump
into the ocean and swim, or stay on the ship. Many jumped to their

RAIN, DRIZZLE & FOG

deaths, thrown against the cliffs by the strong currents or weighed down by leaking fuel oil in the water. Rescuers from St. Lawrence waded into the waves to help survivors to shore and up the ice-covered cliff face. In some cases, victims were too exhausted, numbed with cold, and covered with oil from the sinking ship to even stand; rescuers carried them up over the cliffs on their backs. St. Lawrence residents brought the American seamen back to their homes and administered the best care they could.

Minutes after the *Truxtun* hit the rocks, the *Pollux* went aground about 2 kilometres farther west. At Lawn Point, the crew of the *Pollux* waited for daylight aboard their stranded vessel. With 4-metre waves washing over the ship and nothing but a rocky ledge and a steep cliff in sight, the fact that those on board were only about 20 metres from the shore was cold comfort— the water was the only way out. At first light, after several unsuccessful attempts, five men managed to get close to the shore in a motorized whale boat. The boat crashed on the rocks, throwing the men into the icy water, but they made it ashore. They brought a line with them to attempt further rescues, but this was unsuccessful. Just before noon, the ship began to break up, and the captain gave the order to abandon. As the ship broke apart, many sailors jumped into the water, and perished. Twenty-eight made it to safety, helped by the sailors from the whale boat. Three men scaled the cliff in hopes of finding help and were joined later by two others. Fate and luck intervened. A man from St. Lawrence, Lionel Saint, who had seen debris from another ship while at the *Truxtun* site, decided to look off the cliffs near

Lawn Point. Saint led five *Pollux* survivors through three hours of "howling winds and hip deep snow" to reach the Iron Springs Mine in St. Lawrence, where survivors from the *Truxtun* were being sheltered. Back at the *Pollux*, navigator Grindley managed to throw a rescue line from the ship to Alfred Dupuy on shore, who secured the rope so that all the remaining sailors and the captain could make it off.

Exhausted rescuers hurried from the site of the *Truxtun* to the *Pollux*. Eight men from Lawn—Joseph Manning, Alfred Grant, James Manning, Thomas Conners, Robert Jarvis, James Drake, Andrew Edwards, and Martin Edwards (*Evening Telegram*, February 24, reported these last two names as Edward Martin and Edward Roberts)—took horse and slide, axes, and rope and set off for the scene of the disaster.[96] When they arrived and gazed down the 40-metre cliff, they saw men clinging on for dear life. The fearless men from Lawn and St. Lawrence who executed the rescue worked from 5:30 p.m. until 4 a.m. Numb and freezing, they never let up.[97] Like the men from the *Truxtun*, these rescued soldiers and sailors from the *Pollux* were brought back to St. Lawrence, where the women washed and fed them and provided them with warm, dry clothing.

Those who lived through the ordeal credited their survival, as the US Navy put it, to "the tireless efficient and in many cases heroic action of the people of St. Lawrence, Newfoundland."[98] Although many lives were lost in the two shipwrecks, that of Lanier Phillips was changed for the better. Phillips, the US Navy's first African-

American sonar technician, who had grown up in the American Deep South, was well acquainted with prejudice and racism and conditioned to fear white people. Inspired by the kindness and respect shown him by the people of St. Lawrence, he went on to fight the Navy's segregation policies. In later years, he toured the country speaking about his experiences. Memorial University of Newfoundland conferred on Lanier Phillips the honorary degree of doctor of laws in 2008 for his efforts to end racial discrimination, and in 2011, the Newfoundland government awarded him honorary membership into the Order of Newfoundland and Labrador for his work in civil rights in the US.[99]

Of the 233 on the *Pollux* and the 156 on the *Truxtun*, only 186 survived. The remaining men either drowned in the frigid waters or froze to death before they left the ship. Of the 203 who lost their lives, only 180 bodies were recovered. According to the US Navy court of inquiry set up to investigate the events,

> The civilian personnel of the area near the disaster gave unstintingly of their time, labour, homes, food, and personal effects. They are considered primarily responsible for the saving of practically all the survivors of the USS *Pollux* and, through their care of all the survivors of the USS *Pollux* and the USS *Truxtun*, they minimized further loss through exposure.[100]

A WINTER SCENE ON THE NEWFOUNDLAND RAILWAY, DATE UNKNOWN

The Rooms Provincial Archives Division, A 19-61

Like ships, the not-so-speedy trans-Newfoundland trains, known affectionately as the "Newfie Bullet," also faced regular challenges at the hands of the Newfoundland climate. Trains were routinely held up by snowstorms. It was not unusual to read reports of 1 metre or more of snow on the rails between St. John's and Millertown, and even more on the Gaff Topsails, between Kitty's Brook and Millertown, an area rising 61 to 122 metres above the central plateau on the island of Newfoundland,[101] where high winds and heavy accumulations of snow made it the most difficult spot on the railway's route.[102] Push plows attached to the fronts of the trains helped clear snow, but one year the tracks became blocked completely and it took five days to clear the snow.[103]

TRADITIONAL WEATHER LORE

When the moon sets over the beach of Bell Island (up north), bad weather is sure to follow, but if the moon sets up south, there will be good weather for a month.[104]

Perhaps no scientific proof exists for this adage but long-time resident of Portugal Cove-St. Philip's, near St. John's, Mary Thorne, swears that it is both true and reliable.

WEATHER SCIENCE

From January to April, mid-latitude cyclones form along frontal zones and move eastward across North America. These are more likely to develop into major storms during the colder months when a strong contrast exists between the cold and warm air. Systems moving from the southwest can re-develop along the east coast of North America and track toward Newfoundland. The Gulf Stream adds additional energy and moisture to these storms.[105]

1959

A ferocious winter storm started at 5 p.m. on Sunday, February 15, and dumped 56 centimetres of snow on St. John's. Wind speeds of 217 kilometres per hour blew the snow with a velocity that, as the *Daily News* reported, "was unheard of in this part of the world." Cars littered the snow-covered streets, people were stranded, houses lost their roofs, and many buildings were damaged, even destroyed.[106] Some St. John's residents woke up on Monday morning to find snow inside their homes: the wind was so strong that it blew snow in through cracks in the walls and windows and under doorways.

Three days after the storm hit, St. John's mayor Henry G.R. Mews announced that it would take three to four days to open all the streets, and four weeks for traffic to return to its regular pace. He declared a state of emergency, his prime concern being to provide accessible routes for doctors, firefighters, and ambulances in case of emergencies. Residents were warned to take great caution with fires and heaters. In spite of this, several fires did break out and firefighters were forced to reach them by snowmobile. The *Daily News* quoted Mews: "The extent of the work to be done is almost beyond description."[107]

Shirley Lush of Knight Street succumbed to carbon monoxide poisoning in a car parked at the stadium parking lot on Lake Avenue, near Quidi Vidi Lake. She was one of four young people in the car with its engine idling, trying to stay warm while the storm raged outside. The other three survived.[108]

In the small fishing village at the mouth of St. John's harbour, another disaster unfolded. A snowslide in the evening of February 15 buried a house in the Outer Battery trapping 11 members of the Garland family inside until they were dug out by rescuers. At 1 a.m. on February 16, a more lethal second snowslide occurred. Battery residents saw their lights blink and heard a crash, which they described as 50 boxes of dynamite exploding simultaneously.[109] The snow crash took two houses with it and, in turn, these houses crashed into two more. Several other structures in the Outer Battery were smashed, with damages estimated to be hundreds of dollars.

Clarence Wells, whose house had been completely demolished, survived, as did his wife and 16-year-old daughter—although the hot stove that fell on top of the teen burned her badly. Searchers found the body of his son, Theodore, the next day buried under snow and rubble. Charlie Piercey lived through the snowslide that sliced off the second storey of his house, killing his parents and grandmother. He credits his survival to the steel shower on the other side of his bedroom, which took the weight of the collapsed house, and stopped his being crushed. His younger sister also survived, carried across the road by the force of the snow.

One other life was lost that night: 100-year-old Isaiah Dawe, who had recently celebrated his birthday. Nine people were injured in the snowslide. One minute fifteen-year-old Shirley Noseworthy was sitting in her friend's warm kitchen, and the next she was trapped under snow and a collapsed house, where she remained for 12 hours. Local resident Raymond Riche, with the help of other Battery residents, dug through the snow and debris to rescue her. The blizzard and its whipping winds created drifts 7 metres high, hampering the search for survivors and victims.

Local newspapers recounted the harrowing accounts of people trapped under the snow and how rescuers had found them. For 12 hours St. John's residents had neither power nor telephones, so no one knew what was going on at the Battery. City police could not get there until midway through the morning after the slides; their snowmobiles were unable to push through the snow.[110]

The snow stopped all trains, completely blocked roads, shut down the airport, took radio and TV off the air, and left the fire department snowbound. The RCAF loaned extra snow-clearing equipment to the city to help with the cleanup. After the storm, the temperature dropped to -20.6°C. When someone complained about the snow clearing, Mayor Mews replied that "snow clearing squads are working their hearts out to get the city back to normal ... [they have] risen to the occasion magnificently ..."[111] Two babies came into the world that stormy Monday. One mother journeyed to the hospital on a toboggan; the other managed to catch a ride with a fire truck. Outside St. John's, dozens of cars got stuck on the highway, and 70,000 people were without electricity for most of the day.

This storm set a one-day snowfall record for February 15 of 59.4 centimetres, a record that was not broken until April 5, 1999, when 69 centimetres fell in St. John's.[112]

1982

On February 15, 1982, the offshore semisubmersible drilling rig *Ocean Ranger* capsized and sank off Newfoundland on the Grand

Banks, 274 kilometres east of St. John's. Eighty-four men died, 56 from Newfoundland and Labrador, in Canada's worst marine disaster since World War II. The events, recounted in the *Royal Commission into the Ocean Ranger Marine Disaster*, are harrowing.

The storm that sank the *Ocean Ranger* was first identified as a weak disturbance off the coast of Mexico. From there it progressed northward at a speed of 35 knots per hour, but by 12:30 p.m. on February 14 wind speed had increased to 50 knots and the storm was tracking northwest of the drill site. Meteorologists revised their predictions several times. Forecasts—which first predicted maximum wind speeds of 60 knots per hour and sea heights of 7.2 metres—increased in severity. By the afternoon of February 14, the storm was tracking directly for the drill rigs, and meteorologists predicted wind speeds of 90 knots and sea heights of 16 metres by Monday morning. Late Sunday afternoon, February 14, personnel on land received a telephone call from the *Ocean Ranger* saying that the storm had built up quickly and they were unable to continue drilling; they were experiencing heaves of 7 metres and recurring sea spray in the drill-floor area. The *Ocean Ranger* began to shut down drilling operations after 4:30 p.m. It had stopped completely by 6:47 p.m. when the crew was forced to shear off the drill pipe.

The other two drilling rigs in the area had also ceased drilling. Around 7 p.m., a wave struck the *Sedco 706* and dislodged a shed that was welded to the deck of the drill-floor area. It dislodged several pieces of equipment, damaging a beam underneath the main deck. Witnesses estimated the waves to be 25 metres high. The *Zapata Ugland* drilling rig also experienced large waves but did not receive serious damage.

Between 7:30 and 8:00 p.m. on February 14, the Mobil drilling foreman on the *Sedco 706*, John Ursulak, overheard a VHF communication from the *Ocean Ranger*:

> The panel was wet, [the *Ocean Ranger* employee] was working on it and getting shocks off it ... he had the cover off ... that everything was fine and that they [were] picking up glass, mopping up water, tidying up ... all valves on the port side were opening by themselves.

At 8:44 p.m., senior drilling foreman Jack Jacobsen on the *Ocean Ranger* reported the sea height at 15 metres and wind speeds of 90 to 100 knots per hour. By 10 p.m. the sea height approached 20 metres. Communications from the *Ocean Ranger* to onshore personnel seemed to indicate that the workers on the rig were satisfied that everything was functioning normally again. By 1 a.m. on Monday, February 15, however, shore personnel received a call from the rig reporting a list of about 8 to 10 degrees. Ten minutes later, the mayday went out; the *Ocean Ranger* reported it was listing badly. The last communication from the rig came at 1:30 a.m. The medic/radio operator called Mobil's shore base and said that the crew of the *Ocean Ranger* was ordered to lifeboat stations. The Royal Commission concluded that the *Ocean Ranger*'s port light had been broken in the ballast control room, and the electrical system responsible for ballast control malfunctioned, perhaps as a consequence of sea water coming in. The rig began to list, and the crew was unable to correct the problem. Standby vessels rushed to the scene, but by 3:38 a.m. the *Zapata Ugland*'s standby vessel *Nordertor* reported that "the *Ocean Ranger* has disappeared from the radar."

All 84 crew members were lost. Only 22 bodies were recovered.[113]

2003

Badger has experienced several serious floods in the past half-century. The worst were in 1977, 1983, and, perhaps the most infamous, February 25, 2003. After a heavier-than-usual rainfall in February and mild temperatures, all three rivers—the Badger, Little Red Indian, and Exploits—overflowed their banks with virtually no warning, forming a massive ice jam in the river which forced water out and into the town. People were caught off guard and many escaped, clad only in their pyjamas. Some were trapped in their houses and had to climb out second-storey windows into the buckets of front-end loaders. When the town declared a state of emergency, nearly the whole population of Badger, almost 1,000 people, moved out to stay with friends and family and at improvised shelters in nearby Grand Falls, 28 kilometres away. The flood waters swamped nearly one-third of the town. Badger was uninhabitable, given the risk of electric shock and the fact that the water and sewage treatment plant was under water and inoperable.

Then came a sudden temperature drop. The flooded town froze solid; ice engulfed everything. As the waters froze, residential oil tanks tilted at dangerous angles, bringing the threat of oil spills and environmental disaster. The thick sheet of ice almost buried fence posts and cars, half-submerged houses and businesses, and turned many streets into frozen canals. Not all was frozen, though: Badger Brook, normally only 25 metres wide, swelled to almost 500 metres wide in some places. In the centre of town, the river, ironically, spilled down River Road and into Beothuk Street, flooding the fire and town hall.[114]

Weeks and months later families were permitted to move back into their ruined properties, and many were faced with irreparable damages. At the height of the disaster, some had time to elevate their appliances and save them from the flood waters, others came home to find their belongings ruined and their children's toys frozen to the floor.

A Canadian Red Cross relief fund raised more than $2 million; smaller groups held impromptu fundraisers, each collecting several hundred to several thousand dollars. Churches, businesses, and individuals from across Canada donated to the relief fund. As a result of the disaster, many families moved away from the most vulnerable areas or elevated their homes as a precaution against future flooding.[115]

In the spring I have counted

one hundred and thirty-six different kinds

of weather inside of four and twenty hours.

MARK TWAIN

SPRIING

SHIP IN THE ICEFIELDS, CA. 1900

City of St. John's Archives

MARCH

Traditionally, March, April, and May are considered spring months, but on the island of Newfoundland, anything is possible, weather wise.

TRADITIONAL WEATHER LORE

If smoke from a fire comes out of the chimney and goes straight up, it will be a fine day. If the smoke curls down, bad weather is coming.[116]

On laundry days, Newfoundland housewives would watch the smoke and decide whether to hang their clothes out to dry. A possible explanation for this is related to humidity. When, for example, an oncoming low pressure system approaches, water particles in the moisture-filled air condense on the particles in the chimney smoke, making them of greater density than the surrounding air. When this happens, smoke might sink or travel horizontally.[117]

Vertical temperature changes in the atmosphere may play a greater role than humidity in the flow of chimney smoke. Usually as a storm approaches, the temperature will increase aloft (above the surface) more quickly than near the surface. Because cold air near the surface is denser than the higher warmer air, it traps the chimney smoke and prevents it from rising. As a result, when a storm approaches, smoke sometimes fans out horizontally or, in some cases, descends.[118]

1613

Crout recorded this entry in his diary in Cupids on March 13 and 15, 1613:

> In the morning the wind at northeast much wind verie thick weather all the day untill night and small rayne this morning was a litle froost ther came into this harbour a verie hudg great Iland of ice. this daie proved verie cold untill it was night with much wind and abowte 4 of the clock yt did blowe a verie great storme at east snowing verie thick untill night and continued so all the night snowing Continually and Freessing.

> In the morning the wind at west litle wind somthing cold but the sune shining verie pleasaunt the after noone somthing warme which did desolve the snowe very much and a very pleasaunt daye this after noone William Glissen did tak some 4 or 5 troutes in our brook hard by the house.

> This night yt did freese somthing hard the wind at west southwest litle wind all the night.[119]

TRADITIONAL WEATHER LORE

A very light frost on the trees in the morning indicates that a mild period is coming soon.[120]

Frost, heavy or light, forms on trees in the morning when night skies are clear with light winds. These conditions occur under high pressure systems. Since weather systems usually move from west to east and the wind blows clockwise around a high pressure system, that system moves away, the wind will turn from a cold westerly to a southerly direction and bring in milder air.[121]

1796

In winter and spring, ice is always a threat to travellers. Here's how Thoresby coped in the spring of 1796:

> ... I went to a place at a little distance [from Brigus] to preach;
> I had to go down a high mountain, and then on a path-way
> which led close by the side of a hill; I was obliged to walk on
> creepers, (two pieces of iron made to fit the feet, having prods
> to pierce the ice to prevent the foot from slipping) the sea
> roared in a tremendous manner under us, which made it very
> frightful, and more so in the situation we were in, for had our
> feet slipt we must have unavoidably tumbled headlong into
> the sea.[122]

It is always a relief when the winter snow begins to melt and soften, he adds:

> As the snow is now much upon the decrease, many of the
> men are going into the woods with their rockets on, (they
> are wooden shoes, or pieces of wood the size of the top of

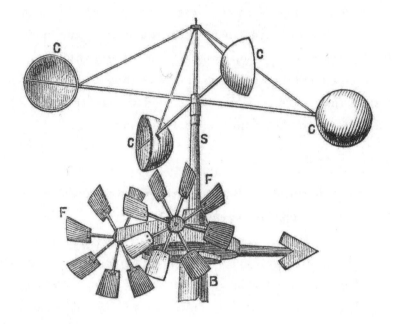

ANEMOMETER

Encylopedia Britannica, 10th ed., 1902

a firkin [small barrel], fastened to the foot to prevent them from sinking in the snow) which are exceeding useful, though very troublesome to travel with ... the men that live in Newfoundland are in general of a hardy race, for many of their houses or tilts are not proof against the weather, numbers of them are open on every side. Several times this winter I have been snowed upon both as I sat in the house and laid in bed; and in some of their houses you might see the men and women, boys and girls, sheep and hogs, hens

and ducks, dogs and cats scrambling in every direction for
to catch a bit of any thing eatable, yet though it is so in
many houses, it is not the case with every one.[123]

Several weeks later, on March 24, 1796, Thoresby made another
entry. Spring seemed a long way off:

> The weather yet continues very severe; it has been frost
> and snow in general ever since last October, and sometimes
> so intense that it has froze the ink in my pocket, nay, it has
> froze the ink in my pen when writing not far from a large
> fire! Those persons in England who have never been here,
> can have no just ideas of the nature of the frost, and the
> depth of snow in Newfoundland in the winter season.[124]

1898

The first official weather observer in Newfoundland, John
Delaney of St. John's, is known as the father of Newfoundland
meteorology. A public servant, politician, postmaster general,
and amateur scientist, Delaney operated several weather
stations on the island between 1856 and 1879.[125] By the late
nineteenth century his assistant, John Higgins, took over the
role—and when Higgins died, his wife took over. In 1898 she
was interviewed by the *Daily News*:

> Yesterday a news representative visited this observatory
> which is located in the residence of Mes-dames Higgins and
> Mitchell on Prescott Street. Mrs. Jno. Higgins is herself the

THE HOUSE OF FRANCIS WILLIAMS, MANAGER OF CAPE COPPER COMPANY, BAIE VERTE, BEFORE AND AFTER THE 1912 AVALANCHE

ASCD

observer and was not a little surprised on being confronted by our Reporter, though she had long expected a visit from a newspaper man. With that thorough knowledge which she possesses of the science, having been the weather observer and reporter for the New York herald and prominent meteorological centres ...[126]

Mrs. Higgins is the first meteorological observer to practice the science by the use of the anemograph which was not here-to-fore introduced. Since the death of her late lamented husband Mrs. Higgins has performed the work in a most systematic and satisfactory manner and during her six years experience has studiously acquired the advance knowledge of anemography.[127]

1912

Tilt Cove, on the Baie Verte Peninsula, became a mining community in 1864, and, by 1869, 800 people lived there. After the Cape Copper Company took over the mine in the 1880s, the community grew to 1,000 people, with its own doctor, telegraph operator, policeman, tailor, blacksmith, and teacher. Most residents built their houses in the snug bowl-shaped cove, but William Cunningham, the telegraph and customs officer; Francis Williams, the mine manager; and Dr. Smith ignored warnings and built their houses under the cliffs at the head of the cove.

On March 11, 1912, a storm raged night and day. Heavy snow, deposited on the ice formed by the freezing rain the previous day, created optimum conditions for an avalanche. The Williamses had just sat down to evening tea when snow came crashing down the cliff. The slide struck two houses, seriously damaging the Williams house and knocking the Cunningham house off its foundation. In the Cunningham house, the maid, Emily Day, was in the kitchen and was buried in the snow along with the Cunningham's three-year-old son. She managed to protect him when she was thrown against the hot stove. Rescuers found the pair; the boy survived, but Emily succumbed to her injuries in hospital in St. John's.

After three hours of digging, rescuers heard voices. Mrs. Williams and her two girls suffered slight injuries, but were alive when the rescuers got to them. Mr. Williams and his 13-year-old son were killed instantly by the collapsing house. The bodies of two of the Williams servants, Peter and Francis Sage, were discovered the next morning. As for Dr. Smith, the snow only grazed his house.[128]

The light of day showed the terrible results of the slide.

1914

The year 1914 was a tragic one for the Newfoundland seal hunt. On March 31, the sealing ship *Southern Cross* was returning home from the hunt, full of seal pelts and riding low in the water when she ran into a storm. The *Portia*, on her way to take shelter in St. Mary's Bay, sighted the ship in the distance through heavy snow and high winds and signalled with its whistle. The *Southern Cross* blew its whistle in return, but that was the last anyone heard from her. Without wireless, no messages could be sent

THE *SOUTHERN CROSS*

City of St. John's Archives

from the ship. For days Newfoundlanders held out hope that the ship and its crew would return unharmed, but the *Southern Cross* was never seen again. All 173 crew and sealers lost their lives. Deemed the greatest loss in the history of Newfoundland sealing, a Commission of Enquiry into the sealing disaster later designated it "an act of God."[129] But according to historian Shannon Ryan in *The Ice Hunters*, however, the disaster may have been an act of man. Captain George Clarke was anxious to win recognition and the prize awarded to the first ship to arrive back in St. John's from the seal hunt. When the *Portia* saw the *Southern Cross* in the distance, the latter seemed to be on her way to Cape Pine, at the bottom tip of the Avalon Peninsula, in apparent disregard of the storm brewing.[130]

On the ice floes off northern Newfoundland, a sleet, rain, and snow storm raged. Men from the SS *Newfoundland* were sealing on the ice when it struck. Due to a combination of arrogance, greed, and bad luck, the sealers endured two nights on the ice, first in a freezing rain storm, and then in heavy snow.

The morning of March 30, 1914, was cloudy with mild temperatures and no sign of inclement weather. Deep in the ice floes, Abram Kean, captain of the *Stephano*, raised his derrick, signalling that there were seals in the area. The SS *Newfoundland*, captained by Westbury Kean, Abram's son, was stuck in the ice 8 to 10 kilometres to the south. Anxious to catch seals, Westbury sent 166 men over the side with instructions to head for his father's ship. The men started out from the *Newfoundland* at about 7 a.m. Thirty-four of the sealers did not like how the sky looked, and turned back. But 132 continued, and reached the *Stephano* by 11:30 a.m.

Captain Abram Kean gave them tea and hard tack and sent them back over the side to hunt seals. He assumed that the *Newfoundland* was only two—not four—hours away. Although it was snowing heavily at the time, he instructed the sealers to kill 1,500 seals before returning to their own ship. On the *Newfoundland*, Westbury Kean assumed that his father would find room for

A SURVIVOR OF THE *NEWFOUNDLAND* SEALING DISASTER IS HELPED ABOARD THE RESCUE VESSEL, APRIL 1914

ASCD

---

them on his ship to wait out the worsening storm. There were no wireless radios on these ships—the company that owned the radios had removed them as their use had not resulted in larger catches, and that was the only reason they believed would justify the expense.

George Tuff, leader of the *Newfoundland*'s sealers, did not question Abram Kean's orders to return to the ice. The men headed back to the *Newfoundland* in knee-high snowdrifts and over unsteady ice pans. When darkness fell, they stopped to build shelters from

MEN ON THE ICE DURING RESCUE AND RECOVERY FOLLOWING THE SS *NEWFOUNDLAND* SEALING DISASTER

ASCD

loose chunks of ice, but this did not protect them from the wind, sleet, and freezing temperatures. Many died that night. Those who survived spent the next day and night trying to get to their ship. Abram Kean assumed they had made it back to the *Newfoundland*, and Westbury Kean assumed they had stayed on the *Stephano*.

It was not until the morning of April 2, after the men had spent two days and two nights on the ice, that Westbury Kean on the

Newfoundland spotted his men through his binoculars, staggering across the ice. He had no flares and no radio, but he improvised a signal to alert other vessels in the area. They rescued the few survivors. Of the 132 men that set out, 77 had died on the ice. A total of 251 sealers died in the two tragedies—the sinking of the *Southern Cross* and the *Newfoundland* sealing disaster.[131]

Newfoundland folklore is filled with tales of second sight and communing with the souls of the dead. Mary Crewe's husband, Reuben, had sworn never to go to the ice again after a particularly difficult voyage, but when the couple's eldest son, Albert John, begged to go, Reuben relented, and accompanied him on the SS *Newfoundland*. As Mary watched them leave, something bothered her. She was sure they were going for the last time. On the night of March 31, when the men were stranded on the ice, she awoke from her sleep with a start, as if someone had touched her. She was convinced she had seen her husband and son kneeling by the bed in prayer. As far as she knew, both men were safe on the sealing vessel. It was not until much later that she found out the grim truth. She was told that Reuben had done his best to keep himself and his son alive, forcing them both to keep walking. When they could go no farther, Reuben wrapped his arms around his son to try to keep him warm and alive. Their bodies had frozen in this embrace,[132] an embrace that inspired Morgan MacDonald's statue commemorating the disaster. The statue is located in Elliston, Newfoundland, the former home of Reuben and Albert John Crewe.[133]

SHIPS IN HEAVY ICE IN ST. JOHN'S HARBOUR

ASCD

WILLIAM TUCKER OF PORTUGAL COVE-ST. PHILIP'S REMEMBERS THE WINTERS OF THE 1930S

Winters seemed to be a lot colder years ago. There was a lot more snow and the weather was a lot harsher. Conception Bay would freeze over when the slob ice would come in around March. Slob was small pieces of ice that just floated in. It froze solid, but I don't imagine it would have been very smooth. You'd see streams of people going over to Bell Island with horses and catamarans picking different paths out, to avoid the rough spots. When it froze, it was strong enough for horses and catamarans to go across. A catamaran was a type of slide. That's how they took the mail, for instance, from St. Philip's to Lance Cove on Bell Island, or to the beach on Bell Island. Even people from Topsail and other places headed over to Bell Island on the ice, because that was quite a mining community at the time. The slob ice would last for probably between a month and five weeks, and then would gradually break up.

According to Tucker, even St. John's harbour froze over, not with slob ice but because of the cold temperatures.[134]

TRADITIONAL WEATHER LORE

Whatever direction the wind is coming from when the seasons change and the sun crosses the "line," that will be the prevailing wind for the next quarter.[135]

If March comes in like a lamb, it will go out like a lion. If it comes in like a lion, it will go out like a lamb.

1958

March 1958 came in like a lion. The *Daily News* headline on Monday, March 3, announced: "Sleet Storm Paralyses Avalon." High winds had started on Friday evening, snapping utility poles like toothpicks, and freezing rain covered everything.

By Saturday, ice buildup on the lines knocked out St. John's power for 12 hours and crippled the city.

Shoppers rushed to Martin-Royal Hardware Co. Ltd. on 159 Water Street to buy blankets and kerosene lamps, and local department stores took the unusual step of opening on a Sunday so that people could buy food, camp stoves, and candles.[136] Bowrings estimated sales of over 2,300 dozen candles, valued at $1,000, that afternoon.[137] Many homes in St. John's had fireplaces and, as the *Daily News* reported, coal sales were high. One supplier sold 49 tons of coal that weekend. People looking for emergency supplies created a lineup outside the Neyle-Soper store on Water Street "equal to the lineup in front of the Board of Liquor control stores around Christmas time."[138]

Even without heat or light, Grace Hospital staff delivered seven healthy babies during the havoc created by the storm. The American Forces at the base in Fort Pepperrell brought a generator to the hospital and by 3 a.m. Sunday morning the building had power again. Several hours later, the Glenbrook Girls Home (a home for unmarried mothers and their babies) called for help. With only one fireplace, the 25 infants there— one of whom was very sick—could not stay. An ambulance transported all of them to the Grace Hospital.[139]

Damage was widespread and, in some cases, spectacular. The 350-foot VOCM tower, located on Kenmount Road, blew

TREES DOUBLE OVER UNDER THE WEIGHT OF GLITTER IN THE GARDENS OF GOVERNMENT HOUSE, ST. JOHN'S, PRE-1918

ASCD

down on Saturday night, taking the station off the air for several hours, and the large steel CBC broadcasting tower in Mount Pearl toppled over with the weight of the ice. The Newfoundland Light and Power Company estimated that it would take at least two months to repair all the damage. Over 500 utility poles were damaged on the Avalon Peninsula, including 50 on the Southern Shore, 27 on Bauline Line, 50 on Thorburn Road, and 15 on Kenmount Road. Mayor Mews asked parents to keep their children at home because of the danger presented by live wires. This warning was not without

**UTILITY WORKERS CUT AWAY ICE-COVERED BRANCHES
NEAR ST. CLARE'S HOSPITAL, LEMARCHANT ROAD**

Daily News, March 3, 1958

reason: one child was seen that Sunday afternoon swinging on a dangling power line, but, fortunately, the line was not live.[140]

The sleet storm hit Bell Island with a vengeance, shutting down the electricity and all access to water. Only one well was accessible for drinking water, and long lines of people carrying pails and buckets resulted. A local café managed to stay open because the owner could cook fish and chips over a coal stove. Mine pumps were without power, and officials estimated that the mines would be closed for 10 days to six weeks. The primary concern, though, was fire. Residents agreed to ring all the church bells in the event of such a disaster.[141] Across the bay, reports from Spaniard's Bay called it the worst electrical blackout since power came to Conception Bay 50 years before.[142] The *Evening Telegram* had so many pictures that on March 3 it published a special "Storm Supplement."

WEATHER SCIENCE

Precipitation usually forms at high altitudes. Because of the low temperatures at that altitude, moisture freezes and forms ice crystals, which turn into snowflakes as they grow heavier, and fall through the atmosphere. They remain as snowflakes as long as the air they are travelling through remains sufficiently cold. If the air temperature warms up as the snowflakes get closer to the ground, they may melt and change to rain. In the case of sleet, the snow melts as it travels through a warm layer of atmosphere

LINING UP FOR SUPPLIES
Daily News, March 3, 1958

but then refreezes as ice pellets if it passes through another layer of cold air. Freezing rain, on the other hand, forms when frozen precipitation in the upper atmosphere comes into contact with warm air on the way down. Here, it melts, and turns to rain. As it passes through another much colder layer with freezing temperatures close to ground level, it does not have time to refreeze, but instead super-cools. When this super-cooled rain hits the frozen ground or objects near the ground, it flash-freezes, coating roads and electrical wires with ice.[143]

TRADITIONAL WEATHER LORE

Sheila's Brush, a batch of snowy, blustery weather—usually the last snowstorm of the season—comes in the days preceding or following St. Patrick's Day.[144]

Sheila, St. Patrick's wife/sister/mother/housekeeper (the exact relation is unknown), supposedly brushes the winter season away. The day after St. Patrick's Day is known as St. Sheila's Day in Newfoundland. April often brings severe winter storms because air masses are much warmer and can hold more moisture, and hence the possibility for a storm or two as spring arrives.[145]

2005
On March 16, 2005, Newfoundland's east coast experienced a storm surge that pushed waves and tonnes of ice up as high as 10 metres along the Avalon and Baie Verte peninsulas. Damage from the wind storm and sea surge was estimated to be in the millions of dollars. Flatrock's breakwater crumpled under the strain of the pounding 30-foot-high waves. As residents watched, it disappeared under the water one minute, reappeared the next as pieces of concrete with reinforced steel sticking out of them, and finally disintegrated.[146] The storm also destroyed boats, sheds, and fishing gear and flooded roads on both peninsulas.[147]

**DUCKWORTH STREET IN DOWNTOWN ST. JOHN'S AFTER
A SNOWSTORM**

ASCD

APRIL

1613

Crout recorded the following on April 20, 1613:

> in the morning the wind at south litle wind rayning untill 12
> of the clock but verie warme weather this morning our bark
> departed for Harbour de Grace to fatch our cables and ankers
> from the French shipp that was taken by captaine Eastone.
> After noone continued Rayning untill night the night very
> myld Rayning all night.[148]

TRADITIONAL WEATHER LORE

Rheumatic pains in elderly people mean a rain will come soon.[149]

Studies show that individuals with rheumatism, rheumatic
arthritis, and fibromyalgia joint pain react to changes in
barometric pressure and temperature; these changes exacerbate
their already painful symptoms.[150]

1796

Weather created rough conditions for travellers, as Thoresby aptly
describes on April 29, 1796:

> Mr. Roberts of Portugal-Cove accompanied me to St. John's;
> the road for the length of seven or eight miles was the worst
> I ever travelled, I was frequently up to the middle of the

legs and water and dirt and sometimes up to the knees; for near three miles we had to walk through the snow that was not dissolved, and sometimes I slipt or fell down up to the middle ... he brought me safe in three hours and a half to St. John's where I was kindly received by the Rev. Mr. Jones: after a little conversation with him, I went to rest weary and faint, but in peace.[151]

TRADITIONAL WEATHER LORE

If the first three days in April be foggy, rain in June will make the grass boggy.[152]

1906

On April 7, 1906, snow fell in Greenspond, Newfoundland, and, according to a 1937 broadcast of *The Barrelman*, hosted by Joey Smallwood, it almost buried the community. William Burry, a guest on the broadcast, said that the snow began to fall on Saturday and, by Sunday morning, it was halfway to the top of the windows on the ground floor of his home. By 4 p.m. it had reached halfway to the top of the house and it was so dark inside the family had to light lamps. Snow continued to fall; at about 10 p.m. the family decided they should get out before the house was completely covered. They escaped safely through a second-storey window. The snow continued all day Monday and into Tuesday. It took 50 men to shovel the house out after the storm. That year the snow in Greenspond did not melt completely until the end of July.[153]

A HORSE AND SLED HAULING LOGS IN CENTRAL NEWFOUNDLAND

Grand Falls Photograph Collection, MHA

1907

The *Evening Telegram* reported that the "express" that left St. John's at 6 p.m. on April 7, 1907, for Placentia only made it as far as Kelligrews and had to wait out the night there. At Lance Cove, Conception Bay South, the sea washed over the railway tracks, making it impossible for the train to proceed.[154]

The storm was the worst anyone had seen in 40 years. It began at 4 p.m. with the wind from the northeast and brought "thick and violent" snow drifting. In St. John's, many people stayed up all night fearing that their roofs would be torn off. Glass flew out of

SNOWSHOEING ON THE HOUSETOPS, POSSIBLY ON MERRYMEETING ROAD, ST. JOHN'S, CA. 1925

The Rooms Provincial Archives Division, A 43-10

---

windows and bricks toppled off chimneys. The *Telegram* reporter compared the storm to that of April 16, 1868, when country houses and fences were buried by snow, and the snowbanks on Water Street reached second-storey windows.[155]

The storm also buried central Newfoundland. Reverend Robert Smith, who served the New Bay / Leading Tickles area between 1906 and 1918, seemed impressed: "APRIL 7, 1907 SUNDAY: What a day! I never saw such a storm of snow and wind. Huge drifts of snow made walking impossible. No one could get out. No church service possible."[156]

Newfoundland-born author Gary Saunders describes the snow accumulation in another early-twentieth-century storm:

> The snow was so deep a horse belonging to E.H. Horwood lumber company perished in its log stable before teamsters could dig it out. The drifts were so deep along the company's 8km / 5mi. portage round that it took five days of hard shoveling to reach my home community, Clarkes Head in Gander Bay. He told me that one of the shovellers laid a plank across the diggings and stood on it as huge Nova Scotia work horses [Clydesdales] walked under it without touching their harness hames [part of the horses' harness that extends above the top of the collar].[157]

1984

Two hundred thousand people on the Avalon Peninsula were left without heat and light from April 13 to 15 after a sleet storm that coated electrical wires with 15 centimetres of ice. Bell Island was without power for a week and the western Avalon Peninsula as well as the communities of Fogo, Bell Island, Carbonear, Upper Island Cove, St. John's, Bay Roberts, Mt. Pearl, Torbay, Harbour Main, and Sunnyside received major damage.[158]

CROCUSES BLOOMING THROUGH MAY SNOW

Photo by Sue Willis

MAY

1613

In this May 9, 1613, report, Crout wrote as if spring had arrived:

> In the morninge the wind at west and so continued untill
> night very faire weather the sune shinginge verie faire all
> the day this day came one Master Rice of Bristoll who
> fisheth at Bellill he brought us a hogghed of beer which
> he bestowed one us[.]

> This night very myld and litle wind at southwest[.]

But, his optimism was dashed the next day, May 10, 1613:
"abowte 9 of the clocke yt did begine to snowe very thicke
untill yt was 4 of the clock in the after noone."[159]

1779

Thoresby wrote about a ship wrecked by ice:

> Trinity harbor, in the month of May, 1779, at that time
> there was large quantities of ice in the Bay, which made it
> very perilous; a bed of ice being just a head of them, Mr.
> Adams bid the man at the helm to turn to one side, but
> through fright or mistake, he ran the brig directly against
> the ice which proved fatal to the ship, the cargo, and Mr.
> William Smith the mate; the rest providentially escaped,
> as we say, with the skin of their teeth.[160]

1791

At Okak, Labrador, Moravian missionaries wrote on May 17 that "the snow blown from the mountains lay 20 feet high around the houses and reached the roof of the church." The weight of snow on one side of their church had pushed the building so strongly that it leaned to one side.[161]

WEATHER SCIENCE

The water is cold around the island of Newfoundland in late spring, but the air that comes up over the island is warmer, which causes sea fog to form. From April to July, the snow and cold Arctic air gradually transform into sunshine and warmer temperatures. The warmer air can hold more moisture and so, when this warm moist air blows over the cold Labrador current, surrounding eastern Newfoundland, it cools and the water vapour in it condenses to form fog.[162]

Newfoundland is well known around the world for its foggy coastline. In July 2001 the Second International Conference on Fog was held in St. John's. It attracted scientists, bureaucrats, and entrepreneurs from all over the world. One local journalist quipped: "In what probably turned out to be a bitter disappointment for many, the weather was sunny and warm."[163]

TRADITIONAL WEATHER LORE

A low fog in the bay during the early morning indicates a good day.[164]

Clear skies at night cause temperatures to cool (a cloud cover tends to keep heat in). Clear skies and light winds are generally associated with high pressure systems. As air cools overnight, water droplets can form near the surface of the water, creating a shallow fog. As the sun rises, mixing occurs and the fog thickens, only to dissipate shortly afterwards.[165]

1880

Harper's Weekly in March 1880 contained this description:

> At certain seasons of the year the fog hangs so heavily over the [Grand] Banks that the approach of the great steamers plying between Europe and America can be perceived only when they are right upon the fleet of fishing-boats and then ensues a lively scramble to get out of their way. Sometimes warning comes too late for some luckless fisherman, who is fortunate if the huge vessel which has run down his frail craft can pick him up and carry him to port. Scarcely a season passes that some casualty of this kind is not reported and it is probably that many a fishing craft that is never heard from after leaving port has come to an end in this way.[166]

1897

In 1897, Gerhard Schott, in *Die Annalen der Hydrographie &*
Maritimen Meteorologie, stated that "Nowhere on this Earth
is there such an area as the Grand Banks where there are
so many foggy areas, a combination of ice and fog that is a
constant danger to shipping almost the whole year round."[167]

1959

Spring weather can be a long time coming in Newfoundland.
William Tucker recalls a spring storm in May:

> I can remember certain storms, like there was one in May
> 10, 1959. My wife and I, we had one child at the time, we
> went over to see my sister, so we drove over [to Bell Island]
> on Saturday. I had a Buick convertible at the time and when
> we went over there, the car was a bit dirty, so we went in
> somewhere around the mines, five or six of us, and washed
> my car and came out. Fortunately that night I put the top
> up. I can't recall if it was really stormy that night, but the
> next morning we got up to see a beautiful day. But all we
> could see of my car was about 3–4 inches of the antenna
> and it was a really high antenna.

> Later, my sister and I we went to Mass. It was less than a
> quarter of a mile away, but slow going as we had to walk on
> a pathway that people had beaten down through the deep
> snow. There was no traffic, nothing moving at all. At the

ON THE GRAND BANKS

ASCD

time I was in the Army and I was supposed to be back over
[to St. John's] by Monday. Well on Sunday we were going
to come back, because on Sunday, May 10 was my mother's
birthday, and that's the reason why I remember it so well,
but I didn't get back until that Wednesday. They had a path
cut through the snow so I could get a taxi. I had to leave my
car there. In Portugal Cove I managed to get another taxi.
We got as far as Gladney's on Portugal Cove Road, but the
driver had to stop and wait for cars to come through from St.
John's. It was a like a tunnel with 12–15 feet of snow piled on
either side. They couldn't open the road with snow ploughs
they had to get the big Caterpillars to do it first.[168]

2001

In 2001, the snow stopped falling in May. A total of 648.4
centimetres of snow (over 6 metres) fell on St. John's that
winter, breaking all previous records and making it the highest
total snowfall in over 130 years of record keeping and the
highest all-time snowfall among major Canadian cities. In the
six-month period between November 1 and April 30, "37 days
recorded at least 10 mm of precipitation. Of these, 25 days had
at least 10 cm of snow, including 5 days with 20 to 30 cm (three
of these occurred consecutively on April 1/2/3) and one event
with more than 30 cm of snow (Dec. 1, 37.6 cm)." Schools and
businesses closed, and snow-clearing costs soared.[169]

St. John's meteorologist Bruce Whiffen wrote "Weatherwise," a popular column in the *Telegram* in the 1990s and early 2000s. According to Whiffen, an average St. John's winter sees an accumulated snow depth of about 20 centimetres. Even though substantial amounts of snow might fall, it melts regularly with the rain or fog. The winter of 2001 brought 180 centimetres of accumulated snow, with little rainfall and almost no drizzle or fog to melt it.[170]

TRADITIONAL WEATHER LORE

If the opposite shore of the bay looks nearer than usual, it is a sign of bad weather.[171]

This adage is associated with temperature profile: usually colder air is near the surface and warmer air aloft before a storm. The colder air creates a lens-like effect near the surface, making objects appear closer than they are. This, however, is not necessarily always a sign of a storm; sometimes this effect will be caused by cold water, which chills the air above it.[172]

When sun rays crown thy pine clad hills,

And summer spreads her hand,

When silvern voices tune thy rills,

We love thee, smiling land.

SIR CAVENDISH BOYLE, 1902

SUMMER

CAPLIN

Photo by Craig Purchase, Memorial University of Newfoundland

JUNE

1583

The first description of Newfoundland climate comes from 1583. Edward Hayes, who sailed with Sir Humphrey Gilbert, wrote that Newfoundland's summer was hotter than that in England: "as in the months of June, July, August and September, the heat is somewhat more than in England at those seasons."[173] Perhaps this was just wishful thinking, or simply what he expected to find, given the common belief of the time that climate was a constant on any given latitude.

TRADITIONAL WEATHER LORE

Rain, drizzle, and fog bring the caplin in.

June is traditionally the month of fog and caplin in eastern Newfoundland. Known in Newfoundland as "caplin weather," a combination of rain, drizzle, and fog seems to invite the little fish onto shore to spawn on the rocky beaches. Some are lucky, and reproduce; others are scooped up by hungry Newfoundlanders—a rite of spring on the island. In addition to caplin, icebergs also arrive in May and June. As Parmenius described in 1583:

> Some of our company report that, in May, they were
> sometimes kept in, with such huge yce, for 16. whole dayes
> together, as that the Islands thereof were threescore [60]

fathoms thicke, the sides whereof which were towards the
Sunne, when they were melted, the whole masse or heape was
so inverted or turned in maner of balancing, that that part
which was before downward rose upward, to the great perill
of those that are neere them.[174]

1767

June can bring every kind of weather, as the Lesters recorded in
their diaries. Isaac and Benjamin Lester were partners in a trading
house in Poole, Dorset, engaged in the transatlantic Poole-
Newfoundland cod fishery and trade. The brothers kept business
diaries on their firm's activities from 1761 to 1802. Included
with the record of the number of quintals of fish was interesting
weather information. Benjamin had left Poole and come to
Newfoundland in 1738 to work for his father-in-law. By the 1750s
he had established himself as a planter and merchant in Trinity,
Conception Bay. He recorded the following on June 6 and 7, 1767:

> Saturday June 6th Strong Gale at NNW under fore Sail all
> Night standing to & from the Horse Chops & So [south]
> Head Catalina Noons Strong Gale Drove of about 2 Leags Sot
> dowble Reef'd Main Sail & fore Staysail

> to get in with the Land, very Cold, Snow in patches on The
> Shoar, not melted since the Winter, in the Evening moderated
> & fine Sky hope it will be at SWt [southwest] in the Night at
> 8 a Clock within half a Mile of So Head Catalina.

A POSTCARD OF A MAJESTIC ICEBERG OFF THE NEWFOUNDLAND COAST, CA. 1900

Newfoundland Collection, A.C. Hunter Public Library, St. John's

---

[Sunday June 7] Morning hazy & sometimes drizling Rain, Wind freshen'd up again at North ...[175]

1796

Impressed by the unique shapes of the icebergs that floated into Newfoundland waters, Thoresby wrote on June 7:

I was greatly delighted with the prospect I was favoured with of several islands of ice forming different figures in a variety of directions; some of them had a very striking appearance of the old Abbeys in Yorkshire, and one of them was as large as St. Paul's in London. This is the first summer's evening we have had this year.[176]

He was, however, worried about the storm two days later: "June 9th, 1796—There hath been much thunder and lightening [sic] most of the day, the thunder was the most awful I ever heard."[177]

TRADITIONAL WEATHER LORE

When the ditch and pond offend the nose, then look for rain and stormy blows.[178]

Gas produced by decaying matter in ditches or stagnant ponds produces gas, which accumulates in bubbles that sit just under the mud. As the atmospheric pressure decreases, when poor weather is on the way, the pressure on the bubbles lessens and they expand and break, producing a rank odour.[179]

1799

At the Nain Moravian mission residents were disappointed with the weather:

The summer was mostly harsh and like winter. On June 10, a quarter of an ell [approximately 58 centimetres] of snow fell, and the bay near Nain became totally free of ice only at the

end of this month. Nearly all potatoes froze. In Okkak the
snow started to melt in the country, yet shortly afterwards
it snowed again and on the 24th two sledges with Eskimos
arrived on the ice. At the end of June, it froze again so strongly
that the whole bay near Okkak was covered in one night by a
thin crust of ice. Also many garden plants froze and the largest
part that remained have been eaten by mice or birds.

By August the sea was still covered in ice.[180]

1874

At the end of the nineteenth century, James Ryan was one of
the most prosperous merchants in Newfoundland, based in the
Ryan Premises in Bonavista. Originally from Ireland, James
Ryan's father, Michael, had started business in Newfoundland by
selling fishing supplies and alcohol (a pub was part of the initial
operations). The Ryan Company became "a dominant player in
the salt cod trade well into the 20th century."[181]

The following excerpt was taken from one of the Ryan diaries,
which were kept from 1874 to 1919 by James Ryan, his office
staff, and, in later years, Nicholas Ryan, James's younger
brother. On June 2, 1874, Robert Brown, the Ryan Company's
bookkeeper, recorded "shocking" weather:

Wind N.E. by E. blowing a stiff breeze causing sea to run too
high for the *F. Cloud* or *Maggie* to proceed to St. John's. Last

night and this morning shocking weather was experienced. It did not cease snowing the whole night and at present (12 A.M.), the ground is covered with snow almost everywhere. Notified Abraham Skiffington at 3 1/2 P.M. that his Boat was riding on my mooring ground and that if she done my boat any injury I would hold him accountable.

8 P.M. Schrs. *Flying Cloud* & *Maggie* both left, the wind having gone further north.[182]

1876

Two years later, summer came early in the Bonavista area, according to the Ryan diaries:

June 14, 1876, Wednesday
Very fine day. heat intense. Wind W.S.W.
Ther. 2 P.M. 78° fah. in the shade -107° in the sun.
Flying Cloud left for King's Co. early, returning with salt this evening.

June 15, 1876, Thursday
Very fine & warm. Wind West.
Lynch arrived Noon—no letters.
Swift with load of ballast sent to King's Co. this morning.
2 P.M. *Leopard* called in from Nor'ard.
4 P.M. Abbott arrived with 50 tubs Herrings hauled at Gambo. Hicks discharged similar quantity 2 hours later.[183]

A FOREST FIRE IN THE INTERIOR OF LABRADOR NEAR THE JACK FISH LAKE (MULLIGAN) AREA, JUNE 2008

Photo by Eric Earle, Department of Natural Resources, Government of Newfoundland

CUMULUS CLOUDS ON A FINE SUMMER'S DAY

Environment Canada

The fine summer weather must have been enjoyable, but it often led to drought, and, with drought, the danger of forest fires. Even though June can be cold and wet on the east coast of the island, central Newfoundland, as well as the west coast and Labrador, is more likely to have hot weather with temperatures in the mid to high 20s. With hot dry weather comes the danger of forest fires, some started by man, but others by natural sources. Thunder and lightning storms have historically started some of the province's worst forest fires.

1888

Nicholas Ryan writes from Bonavista on Wednesday, June 6: "Fine Beautiful day. Forest fires are raging at Bird island cove and also at King's cove (Bonavista Bay)."[184]

WEATHER SCIENCE

Fluffy cumulus clouds are often seen on fine summer days in Newfoundland. They are known as fair-weather clouds, and they form in the day but dissipate at night. If they grow in height, they may develop into towering cumulus which, if the atmosphere becomes unstable, may become convective cumulonimbus or thunderstorm clouds. Some areas of Labrador and central Newfoundland are more prone to thunderstorms.[185]

TRADITIONAL WEATHER LORE

*When the wind is in the south, it blows the bait in the fishes' mouth. /
When the wind is in the east it's good for neither man nor beast.*

According to Patrick Coish, who fished for years out of the
Battery in St. John's, the "old man" (captain) would not go out
at all if the wind was from the northeast, because there would
be no fish in the traps. But if the wind was from the southwest,
the fishermen always had successful voyages.[186]

1961

The summer of 1961 was one of the worst in the memory
of many Newfoundlanders. Forest fires blazed over much of
the island. Although some of the fires were caused by human
error—campfires, careless smokers, and sparks from the
railway—the Newfoundland winds increased the damage.
Over 265,000 hectares of forest land burned.[187]

One of the most serious fires burned over 134,885 hectares
of productive forest land in Bonavista North and Fogo.[188]
Gander, Trinity, Carbonear, Bay de Verde, Port de Grave, Fortune
Bay, Ferryland, and St. John's also experienced forest fires.
Between August 24 and September 22, 1,080 Royal Canadian
Army troops assisted local firefighters. Most of the damage
to productive forest lands occurred on private holdings, not
crown property. About 36 homes were also destroyed.

One of the largest of the summer fires on the island occurred in Clarenville. It started as a small fire near Dunn's River on the Burin Peninsula 60 kilometres from Clarenville, but, aided by strong winds, it quickly travelled to Clarenville. Local authorities were ready to evacuate the population by train but, just in time, the wind changed direction and the fire moved off toward Port Blandford. A Canadian Army unit known as the Van Doos, along with local volunteers, successfully battled the flames.[189]

1997

It can, however, get chilly in June. On June 24, 1997, Queen Elizabeth II visited Bonavista to celebrate the arrival of the *Matthew*, a replica of John Cabot's ship. She was greeted by a temperature, with the wind chill, of -2°C,[190] and used a red plaid blanket over her lap to keep warm.[191]

RENNIE'S RIVER SWIMMING POOL

City of St. John's Archives

JULY

When July comes to Newfoundland, residents can *usually* relax and enjoy the summer sun. After the long winter and wet, foggy spring, July almost always brings the long-awaited summer weather.

1767

The Lester diary, Trinity, recorded on Thursday, July 16:

> Morng Calm & very hot Weather at 10 a Breeze at SEt put
> 3 Scifts Load Capling on board Capt Riders Boat, ... got the
> Meat out of the Bee, & Scrubb'd her Bottom, Mr Gaulton
> came in, after being out since Munday Morng, & only brought
> one Quintal fish

> Fryday July 17th
> Morng little Breeze at S.E. Capt Lemon Ship came in at 4
> a Clock, 13 Days from Philladelphia, Sent the Lighter to get
> a Load Ballast to goe on board the Bee ...[192]

1791—96

The following excerpts are from the Okak Moravian weather diary.

> July 2, 1791: The weather in June was very wintry and even on
> 2 July, the Eskimos caught 5 seals on the ice, and still rode

with sledges around on it. The next day, the ice broke up, and on 5 July, the first kayak was launched.

Mid-July 1792: It was very warm here [Okak]. The Fahrenheit thermometer rose to 92 degrees.

July 1796: By the end of July, one did not know how to hide from the great heat ... by the end of August the summer [in Labrador] closed at once with a heavy thunderstorm. From this time the winter was changeable in the months September till December.[193]

WEATHER SCIENCE

A sunshine recorder consists of a glass sphere set in a metal base with a piece of cardboard mounted at the back. The cardboard is marked off in hours, with the local noon in the centre. The amount of hours of sunshine in a day is shown by the burning or scorching (depending upon the strength of the sun) the graduated piece of cardboard. Known as a Campbell-Stokes recorder or a "Stokes sphere," it was invented by John Frances Campbell in 1853 and later modified by Sir George Gabriel Stokes in 1879.

TRADITIONAL WEATHER LORE

When it is going to rain, the trees, especially spruce, will turn a darker green.[194]

**A SUNSHINE RECORDER SITS NEAR THE DESK
OF ST. JOHN'S METEOROLOGIST BRUCE WHIFFEN**

Photo by Bruce Whiffen

Perhaps because the sky will probably be overcast before a rainfall
and the light dim, the trees may appear a darker colour.[195]

WEATHER SCIENCE

High, white, wispy cirrus clouds are a common sight on a
fine summer's day. However, if they thicken and are followed
by progressively lower, more continuous cloud, a change of
weather could be on the way—a warm front associated with an

approaching low pressure system. The direction in which the streaks are moving gives an idea of where the next batch of weather may originate.[196]

1816

In April 1815, a volcano on the island of Sumbawa, Indonesia, erupted with such force that the explosion was heard 1,400 kilometres away. The largest volcanic eruption in recorded history, over 92,000[197] people died from its effects or as victims of the starvation and disease that followed. In the Indonesian province of Tambora, only 26 out of a population of 12,000 survived.[198] Hot ash shot 43 kilometres into the air released sulfur dioxide and water into the stratosphere, combining to make sulfuric acid aerosols, which formed a heavy cloud cover that reduced solar radiation to the earth and disrupted weather patterns worldwide. Temperatures dropped and rainfall patterns changed; dry areas became wet, and wet dry. Because these aerosols formed so high in the stratosphere, they stayed there for years. Weather data for that time period indicates there was a two- to three-year period of extreme weather after the eruption. In 1816 and 1817, the effects of the volcano's eruption were felt in Europe and North America and caused what came to be known as the year without a summer:

> The persistent cold of 1816 triggered increased precipitation and storminess: it weakened atmospheric circulation by reducing the difference in air temperature (and hence air pressure), thus keeping the polar front from retreating north at its normal pace.[199]

CIRRUS CLOUDS OVER TWILLINGATE

Photo by Elizabeth Barney

Average temperatures in the northern hemisphere dropped by 0.4°C to 0.7°C.[200]

Several other volcanoes also erupted in the early 1800s. The combined effects of all these eruptions, which took place toward the end of a longer period of weather disruption known as the little ice age, may have added to the severity of the weather.[201]

Europe was also suffering from the devastation of recent military conflict. With the end of the Napoleonic wars, many farms lay in ruins, and the cold rainy weather created a disastrous growing season. Food shortages abounded and prices skyrocketed. Rioting, pillaging, and violence broke out all over Western Europe.[202] Emigration increased as the situation worsened. Between 1816 and 1818, for example, Irish migration to the United States and Canada rose from 6,500 to 20,000.[203] The immigration to Newfoundland at that time was mainly due to the boom in the fishery. The Slade and Kelson diaries recorded at Trinity in 1816 noted that

> Monday 3rd June 1816—Wind S.W. weather indifferent put in here the Brig Joyce, from Waterford with abt 330 passengers (including 40 women) bound to St. Johns being very short of Water, arrived The Marias, Captn. Hall from Waterford [Ireland].[204]

In the United States, the summer of 1816, or "Eighteen hundred and froze-to-death," had unusually cold weather. Snow and ice pellets fell in New England on June 5, and "wind during the whole day as piercing and cold as it usually is the first of November ..." In Maine, July frost killed beans, cucumbers, and squash. In August the frost struck again, injuring crops in New England, Maine, and New Hampshire.[205] Likewise, in the Maritime provinces of Canada, crops and animals suffered when severe cold killed crops in each of the summer months of 1816. In September, another major frost destroyed any surviving crops. The *Halifax Chronicle* of December 1816 reported:

Great distress prevails in many parishes throughout [Quebec]
Province from a scarcity of food ... many of them have no
bread ... It has been given us from the authentic sources that
several parishes in the interior part of [Quebec] are already so
far in want of provisions as to create the most serious alarms
among the inhabitants.[206]

The year 1816 brought trouble in triplicate to Newfoundland.
Between 1815 and 1818 the economy collapsed; the population
suffered some of the coldest winters ever recorded; and two
devastating fires destroyed the capital city of St. John's. In all
of this trouble, wind, rain, frost, and ice played a leading role.

Coinciding with global weather disruptions worldwide,
Newfoundland's boom years ended at about the same time as
Britain's war with the United States (1814–15). American fish
flooded the European market, and Newfoundland fish prices
plummeted—from 40 to 19 shillings per quintal.[207] In Battle
Harbour, Labrador, the average price dropped from 20 to 12
shillings per quintal in 1816 and, the next year, to 8 shillings.[208]

As a consequence, many merchants in Newfoundland and England
went bankrupt, causing widespread unemployment among the
local population. Because so many businesses failed, not as much
food was imported to the island, and food shortages became a
serious problem. In the 1800s, Newfoundland imported much
of her supplies from Great Britain and North America. In 1807,

21,621 hundredweight (1 hundredweight is 100 pounds) of bread
flour came from Britain, 51,134 hundredweight from the United
States, and 27, 252 hundredweight from British North America
(probably from or via Quebec).[209] The governments of New
Brunswick and Nova Scotia prohibited the export of grains for
several months.[210] This, in addition to heavy sea ice blocking
transportation routes, thwarted attempts to get food into
Newfoundland.

The ice settled around the bays and harbours as early as
November 1816, blocking the entire east and northeast coasts
and further isolating communities in these areas. Labrador fared
even worse; severe sea ice clogged the ocean through the summer
of 1816. A Moravian report records the ubiquity of poor weather:
"As in almost every part of Europe, so in Labrador, the elements
seem to have undergone some revolution during the course
of last summer."[211]

The unusual weather caused by the Tambora eruption brought
a collapse in the fishery worldwide. In Newfoundland, fishing
efforts were thwarted by wet and windy weather; fishermen were
often unable to cure the fish they did catch. Governor Pickmore,
in a letter to the Earl of Bathurst dated September 30, 1816, wrote:

> The Bank Fishery as well as that carried on by Boats has
> been in regard to catch tolerably abundant but owing to a
> prevalence of Rain and bad weather in the early part of the

season much Fish has been spoiled and although it affords at present a better aspect, the Exports of this Year, it may be apprehended, will fall considerably Short of former ones.[212]

Heavy ice prevented a profitable seal hunt again in 1817. Judge Daniel W. Prowse, historian and author of *A History of Newfoundland* (1895), noted that the calamitous streak ended in the spring of 1818 with a productive seal harvest and a lucrative fishery.[213]

In the winters of 1816 and 1817, then, many Newfoundlanders faced a lack of food and money. Making matters worse, two major fires broke out in St. John's, affecting those well beyond the city limits. St. John's extended north to New Gower Street in the east and Carter's and Barter's hills in the west.[214] The fire of February 1816 destroyed 130 dwellings and left 1,500 people homeless. It was reported that a hurricane checked the progress of this fire.[215] The winds associated with the storm probably helped spread it, but the precipitation may have helped to douse it.

The Great Fire of 1817, which was actually two fires, burned 300 homes and left 2,000 people homeless but, perhaps more tragically, it also destroyed the leading provision houses and their winter supplies.[216] With its rough wooden houses built close together, and almost always a breeze of wind to fan any flames, St. John's was a dangerous place to live. It is described this way in an 1809 insurance report:

> The street ... only a crooked and narrow alley, in some places
> contracted to 6 feet wide, in others expanding to as much as
> 12 or 18 feet. This street runs nearly in the direction of the
> prevailing winds, and is a tunnel through which they blow
> occasionally with great violence.[217]

Rumours circulated in St. John's that arson may have been
the cause of these tragedies. As naval officer William Glascock
described, in the years after the governors began overwintering
in Newfoundland, upon the close of the fishing season, and
the return of the predominantly Irish fishing servants into
port, an annual fire "was as regularly looked for as the coming
of the frost"[218]:

> It is whispered among the better informed of this island, that
> some of the mercantile community have most opportunely
> escaped bankruptcy by what might almost be termed a
> providential conflagration.[219]

Despite these suspicions and rumours, a grand jury ruled that the
fires were accidental.

Meantime, hunger reached catastrophic proportions in the
winter of 1817: "now famine, frost and fire combined, like three
avenging furies to scourge the unfortunate Island."[220] A grand
jury document described the desperate acts of several hundred
men without money, food, shelter, or adequate clothing: "gangs of
half-famished, lawless men everywhere threatened the destruction

of life and property." They looted stores and destroyed property not only in St. John's but also in Renews and Bay Bulls. Ships and crews stuck in the ice were easy targets. That winter was known as the "Winter of the Rals" (rals are rowdies or vagabonds; people exhibiting disruptive behaviour).[221]

In St. John's, 800 emigrants were sent back to Europe because of the food shortages.[222]

Organizations such as the Benevolent Irish Society (BIS) helped not only the Irish population of the city but also those of other creeds and nationalities. BIS president James McBraire generously gave his own money to help relieve the distress.[223]

WEATHER SCIENCE

In Newfoundland, wind often accompanies fine weather. Cool air over the ocean rushes in to replace the rising warm air over the land—this process creates an onshore wind or sea breeze. This can cool down a hot day. The opposite can happen as well when cooler air moves from land and out to sea, an offshore wind or land breeze.

Newfoundlanders generally do not have an opportunity to complain about heat and lack of rain, but in the summer of 1911 "The Great Drought" hit Newfoundland. It caused over 1,000 fires—maybe as many as 1,193 fires—some of which were blamed on sparks from trains and the carelessness of berry pickers.[224]

1946

On July 27 the rainfall was on the front page of the *Evening Telegram* right next to the war news. Labelled a "gully washer deluxe," it washed out small bridges and flooded roads, ditches, and homes. A total of 121.2 millimetres fell on St. John's.[225] The rain started shortly after 6 p.m. on Saturday, July 27, 1946, catching homeward-bound city workers. A series of heavy downpours punctuated with respites of drizzle periodically fooled people into thinking the rain had ended: "Pedestrians had reason to believe it was only more examples of the weatherman's evil prankishness when just after they emerged from homes ... down came more tons of rain ..."[226]

It rained all Saturday night and into Sunday morning. As a result of the rain a "miniature Niagara" formed down the embankment near the north bank of Rennie's River swimming pool, a "white-foamed deluge" washing tonnes of clay and stones into the pool. On the Southside Hills, torrents of water gushed down the steep incline of the city's southern boundary. Bowring Park, in the west end, lost several bridges, as the flooding river roared into the valley, sweeping debris into the harbour. At Quidi Vidi Lake, in the city's east end, the river overflowed its banks and covered the road up to the boathouse, forcing the oarsmen to observe an evening of rest (on Saturday). Despite the weather and the flooding, however, they were on the lake on Sunday morning. With Regatta Day so close, a disruption was a concern, but officials reported that the lake would be back to normal on Monday.[227]

RAINSTORM DAMAGE NEAR 355 SOUTHSIDE ROAD IN ST. JOHN'S, POSSIBLY JULY 27, 1946

City of St. John's Archives

Not far from St. John's, the river at St. Philip's increased to many times its normal size, running straight over the beach and into the ocean. The wharf and breakwater had to be rebuilt as "[t]he river rose over the beach and scoured out a passage about nine feet deep."[228]

TRADITIONAL WEATHER LORE

If you can hear a train whistle from a distance, it is a sign of bad weather to come.

A NAP IN THE SUN, CA. 1970

Photo by Sheldon LeGrow, City of St. John's Archives

Considering that Newfoundland has not had a train since 1988, this bit of weather lore is no longer applicable, but perhaps a possible explanation does exist. In summer, sound travels farther in humid air, which also often foretells a coming shower or thunderstorm.[229]

1949

On July 7, 1949, the temperature in St. John's soared to 30.6°C, the hottest day on record for the city.[230] Fortunately, the next day, Bannerman and Victoria parks opened for their 26th season. The Mount Cashel Band played under Reverend Brother Brennan

BATHING BEAUTY AT THE SEASHORE, LOCATION UNKNOWN

City of St. John's Archives

and children gave three cheers as Mayor Carnell gave his opening
remarks. Carnell cautioned the children to be careful when in
traffic while coming to and from the park. "Good kids are scarce,"
he commented. These "good kids" must have been delighted
because both Bannerman and Victoria Park playgrounds received
new equipment costing $3,000, including box car swings, giant
strides, ocean waves, multiple slides, and four board see-saws. The
mayor announced that a new paddling pool would be constructed.
With weather that hot, it would be better sooner than later for
the children of St. John's.[231] Relief from the heat would normally
have been available at the Rennie's River and Tooton Pools

(Victoria Park), which usually opened a few days after the park did. The pool openings, however, were delayed that summer. In the case of the Rennie's River pool, it was possibly because of a necessary cleanup: a storm the previous fall had deposited 500 tonnes of rock and shale on the bottom of pool.[232]

The high temperatures may have been connected to a heat wave searing across the eastern United States, which caused 153 deaths and serious crop losses in Massachusetts, Rhode Island, and Connecticut.[233]

TRADITIONAL WEATHER LORE

When a kettle boils dry on the stove, it's going to rain.[234]

Because of the high amount of rain that falls in Newfoundland each year, there is a good chance of this one being right!

PAUL O'NEILL REMEMBERS SUMMERS IN BAY DE VERDE AND IN ST. JOHN'S

We mostly went down home [to Bay de Verde] in the summers after having spent the winters at school in St. John's. We had complete freedom. None of the kids had summer jobs except for sometimes the poorer kids had jobs carrying fish. We swam and paddled in the swimming hole in Bay de Verde. When we stayed in St. John's, we swam at the Rennie's Mill River at the foot of the waterfall which was a children's pool then.[235]

**FLAKES AND SHEDS WITH HOUSES AND CHURCH IN THE
BACKGROUND, BAY DE VERDE, CA. 1930**

The Rooms Provincial Archives Division, VA 15A-14.1 / C.L.

2001

On July 25, 2001, the Barbour Property in Newtown was struck by lightning for the third time in 10 years. The lightning shattered glass and splintered boards as it hit a beam in the upstairs dinner theatre gallery. With damage to its furnace, sound system, and an elevator, the estimated price tag on the strike was in the thousands of dollars.[236] However, the theatre troupe did not miss a performance or a rehearsal because of the lightning strike. Instead the troupe created a character that walked around wearing a metallic colander on his head, attached by a wire to a long metal pole that he carried in his hand. It achieved the desired effect.[237]

2007

On July 31, 2007, a wicked witch named Chantal flew in from the southeast. This post-tropical storm dropped well over 100 millimetres of rain across eastern Newfoundland—up to 200 millimetres in Argentia[238]—and blew a gale with wind speeds up to 85 kilometres per hour. It left behind approximately $25 million in damages, flooding towns, washing out roads, isolating communities, and destroying property.

A welder on his way from Placentia to St. John's to catch a flight to Fort McMurray received only minor injuries when he drove over a stretch of road that he thought was solid, only to crash down into a 6-foot-deep pit. The culvert below the road had washed out and left a shell of pavement and dirt above it.

Placentia's mayor Bill Hogan described the storm as the worst he had seen in 40 years, forcing the town, and several nearby communities, to declare a state of emergency. He described the flooding in one home as water flowing in through the front door and out the back. Whitbourne received 150 millimetres of rain, closing a section of the Trans-Canada Highway for several hours, and Dunville 90 millimetres, which flooded large sections of roads and forced some residents to use their boats for transportation. Bryant's Cove, Bay Roberts, and Cupids also declared states of emergency. St. John's recorded 96 millimetres of rain, which broke a 1975 record of 80.5 millimetres. Argentia's 200 millimetres of rain fell in 12 hours, doubling the historic rainfall record. In all, Chantal's behaviour surpassed the 100-year storm statistics.[239]

TRADITIONAL WEATHER LORE

Mackerel sky and mares' tails make the sailor furl his sails.

Mackerel skies are cirrocumulus or altocumulus clouds. If these thicken and become more numerous, they often signal a low-pressure system or even warn of a thunder storm.[240]

Mares' tails are cirrus clouds, thin wispy strands in tufts which, like mackerel skies, can thicken in advance of a low pressure system.

SILHOUETTE OF A MAN UNDER A MACKEREL SKY

The Rooms Provincial Archives Division, VA 14-254 / Gustav Anderson

The summer of 1999 was warmer than average throughout
Newfoundland and Labrador, resulting in forest fires and
low water levels.[241] Two years later, the summer also had dry
weather, but because of heavy snowfall of the previous winter,
the outcome was not as severe. Thirteen years later, the
summer of 2012 had so many warm summer days and nights
that it will no doubt go down in history as one of the best
Newfoundlanders have experienced.

A FOREST FIRE IN THE CROOKED RIVER AREA OF LABRADOR, 2008

Photo by Eric Earle, Department of Natural Resources, Government of Newfoundland

AUGUST

TRADITIONAL WEATHER LORE

When a heavy tide is running and your bobber is under water, you are sure to get a storm.[242]

It is possible that, as an intense storm approaches, the sea level rises in response to the low atmospheric pressure associated with that weather system. These higher water levels sometimes come ashore ahead of the storm as a heavy or rising tide.[243]

1583

Parmenius described the weather on August 6, 1583:

> The weather is so hot this time of the yeere, that except the
> very fish, which is layd out to be dryed by the sunne, be every
> day turned, it cannot possibly be preserved from burning ...
> The ayre vpon land is indifferent cleare, but at Sea towards
> the East there is nothing els but perpetuall mists, and in the
> Sea it selfe, about the Banke (for so tey call the place where
> they find ground fourty leagues distant from the shore, and
> where they beginne to fish) there is no day without raine[.][244]

1790

Moravian missionaries recorded these weather conditions for
August 2:

The Bay of Nain was only free of ice in the beginning of August. On 2 August there was a lot of lightning in the night. The Eskimos for whom this was somewhat strange awoke the missionaries as they believed that the house was on fire.[245]

1873

Over August 25 to 26, 1873, the first newspaper reports of one of the great Newfoundland "August gales" were published. The storm came, as many Newfoundland storms do, from south in the Atlantic Ocean and made its way up to Nova Scotia as a category 3 hurricane. It destroyed boats, homes, and businesses to an estimated cost of, in 2014 currency, about $70 million.[246] On its way to Newfoundland, with peak winds of 185 kilometres per hour, the storm foundered 1,200 vessels and killed an estimated 598 mariners. Shortly after it arrived in Newfoundland, it weakened to a category 1 hurricane, but it still packed a vicious punch.[247] As it blew into Newfoundland's Avalon Peninsula, it took the lives of 100 people.[248]

1876

On August 29, 1876, a rainstorm poured 173.2 millimetres of rain on St. John's. It broke all records, making it the greatest single daily accumulation ever recorded in the province.[249]

1883

Exactly 10 years later, on August 26, 1883, a "cyclone" hit the Grand Banks: "In the whole experience of the fishing fleets that resort to the Great Banks in Summer and Autumn for fishing

purpose, no more destructive gale has swept over these prolific fish meadows than the cyclone of Sunday, the 26th of August ..."
It annihilated a large portion of the St. Pierre fishing fleet. Every ship lost men, dories, deck gear, trawls, hawsers, or anchors.
Frank Leslie's Illustrated News of October 20, 1883, reported that "The French fishing vessel, *Helene*, left two men and two dories to appease the storm." When the cyclone hit, thousands of fishermen were in their dories checking their trawls and away from their vessels. According to one captain, "it was a miracle so many escaped," one the report attributed to solid ship construction. The old-fashioned French ships, able to carry 100 to 400 tonnes cargo, square-rigged and heavy aloft, tended to roll and labour heavily in high seas, while the American, Nova Scotian, and Newfoundland vessels were "light, buoyant schooners ... [which] ride out a storm at their moorings with wonderful ease and safety. The quantity of cod fish lost in this storm was estimated to be about 40,000 quintals ..."[250]

TRADITIONAL WEATHER LORE

To see a rainbow in the sky is a sign there will be no more rain that day.[251]

This is probably wishful thinking.

1889

On August 14, 1889, enormous hailstones hit the Greenspond, Bonavista Bay, area. The morning of the storm opened with light and variable wind conditions, and a barometer reading of

29.90 kilopascals. At 12:30 p.m., the wind suddenly veered to the southwest. A thunderstorm rolled in with crashing claps of thunder and brilliant lightning flashes and flung down hailstones measuring up to 6 centimetres in circumference. The potato stalks in the gardens broke off, and several windows in the community were shattered. A heavy rainfall followed the hail and drenched the area, and the wind changed from southwest to northwest before it cleared. The *Evening Telegram* reported that the atmosphere was foggy and warm after the rainfall and not fit to handle fish.[253]

1890

From 1890 to 1896, St. John's geologist and surveyor James P. Howley explored the interior of Newfoundland. An excerpt from his diary of August 2, 1890, states:

> The men, on returning with their last packs, reported fearful forest fires away to the south and east in the Exploits valley. This is an awful pity and I fear as the woods and ground are as dry as tinder and the wind high it will make sad havoc with the grand forest of the Exploits valley ... I think it very probable the lightening [sic] yesterday may have ignited the dry moss.[252]

1898

The 1898 Ryan diary contained this observation for Friday, August 5: "A dense smoke is over the town [Bonavista] caused by the forest fires that rage in Bonavista Bay." The entry also reported on the state of cod and the catch of the day.[254]

WEATHER SCIENCE

Most low pressure systems are spread over a large geographical area and announce their arrival with weak easterly winds well in advance of the stronger winds. Hurricanes, unlike these other systems, are usually concentrated in a relatively small area. The barometer may not drop, and the winds may not increase until a few hours before the hurricane hits.[255]

Twisters have been reported in Newfoundland. In Trinity a tornado moved a house off its foundations, knocking down a boat shed and bending over trees. Witnesses say that the sky turned black, the rain came in torrents, and hailstones fell. When the tornado roared across the open water, it pushed the water into a wall about 30.5 metres high.[256] In 1992 a tornado tore through Bishop's Falls, and in 2007 "something" ripped through Whelan's Lake on the Bay d'Espoir highway. Meteorologist Herb Toms of the Gander weather office said there was a "strong possibility the disturbance was a tornado." Whatever it was, it damaged local cabins and boats and a 40-foot wharf. Observers saw a funnel cloud and heard a sound like a freight train.[257]

1927

Newfoundland and Labrador is often walloped by gales in August. One of the most infamous storms to hit the province occurred on August 25, 1927. This storm was the first of two gales to take lives, destroy property, wash out roads and bridges, flood houses, sink boats, and wreak havoc on the people of Newfoundland and

Labrador. The 1927 gale is ranked as one of the top three tropical cyclones in terms of lives lost in Canada since 1899. A category 2 hurricane, it struck Nova Scotia with wind speeds of 166 kilometres per hour. In Curling, in the Bay of Islands, wind speeds reached 145 kilometres per hour. Most of the deaths in Newfoundland as a result of this storm occurred at sea when ships were damaged or lost. The estimated death toll was between 156 and 171. More than 40 were fishermen from the Placentia Bay area.[258]

Families in Nova Scotia received a monthly compensation for the damages caused by the storm. Nova Scotian widows received $30 a month for the rest of their lives and, if they remarried, $7.50 a month for each child under 16.[259] Newfoundland families did not receive anything until an Anglican minister advocated on behalf of the families and obtained $100 a year in compensation for those that suffered the greatest losses.[260]

The *Evening Telegram*'s August 27 report on the storm concluded with the reporter's attempt to cheer up his readers and, no doubt, to support the Anglican minister's petition:

> We feel certain that with the dogged resolution characteristic of Newfoundland fishermen, they will not permit disasters even of such a serious nature as this to defeat them, but they are deserving sympathy and help, and it is to be hoped that some measure may be found whereby they may be enabled to prosecute their calling while the opportunity still offers.

Amidst the reports of death and destruction came the sad tale of the *Annie Healey*, reputed to be the best fishing vessel in Placentia Bay. When a storm suddenly arose, Captain John Mullins and his crew—Michael Mullins, James King, John Foley, Charles Sampson, Patrick Bruce, and John Kelly—did the sensible thing, and headed for shore. Without warning, the seas became extremely high and the winds picked up speed. When the vessel was several miles offshore from Fox Harbour, it turned over, and all the men were lost. When John Kelly's wife, Ellen, heard that her husband had drowned, she said she already knew. She had heard someone throwing lumber around near the back of the house and knew that it was a "token," a sign of his death. John Kelly was only 37; he and Ellen had seven children.[261]

In another bizarre incident, fishermen out on the water after the storm spotted a ship far out to sea with something hanging from the crosspieces near the top of the mast. When they arrived at the ship, they saw a man dangling by his right arm, dead but still holding on. When the ship rolled, he dipped into the sea. John C. Loughlin, the skipper of the ship, had clung on even in death. The fishermen retrieved his body, the only one left on the ship, and brought him home for burial.[262]

1935

The gale of 1935 blew in on August 24. It blasted up the east coast of North America, passing Florida, Georgia, and the Carolinas but then veered out to sea, lessening concerns that it would cause more damage on land. As the storm moved over the ocean, it seemed to lose strength but it came up against a large trough of low pressure as it passed Nova Scotia. The difference in temperatures caused the storm to ramp up again. The collision also changed the storm's direction, throwing it into a course headed straight for Newfoundland. The gale struck with full force late in the evening of August 24, and continued until it blew itself out on the evening of August 25.

News of the full extent of the damage made it home to the families in agonizingly slow motion. Reports of capsized, wrecked, and missing vessels from all over the province were carried by person, ship, and cable to the recipients of the news. "Schooner bottom-up," "many schooners reported ashore," "considerable property damage," and "loss of life" were phrases sprawled across local newspaper pages. The *Evening Telegram* claimed that it was the worst storm to hit the coast in 36 years. Roads and bridges were washed away and a bungalow under construction on Newtown Road in St. John's shifted 2 to 2.4 metres off its foundations with the force of the wind. The storm blew down fences, ruined crops, sank motorboats, collapsed stages, and washed away fish. The sustained wind velocity for this storm was reported to be 59 kilometres per hour, with gusts strong enough to blow telegraph poles and trees out of the ground.[263]

The Walsh family from Marystown suffered a particularly keen loss. Two of their three schooners out that night were lost. The day after the gale, the Canadian government sent the SS *Argyle* and the SS *Malakoff* to look for survivors. During the search, a dory from the *Annie Anita*, Captain Patrick Walsh's boat, was found. St. Shott's residents found the wreck of the *Annie Anita* on the beach at nearby Hazel Cove, broken in two. When they went aboard, they discovered the captain's body in the cabin, and one of his young sons dressed in his best Sunday suit and still clutching a Bible. There was no sign of the rest of the crew.

Fishermen found the *Mary Bernice*, the second Walsh vessel, bottom up near the Virgin Rocks off Placentia. They towed it ashore, but found no bodies on it. The 1935 newspapers reported that over 25 lives from the Marystown area were lost in this storm.[264] Environment Canada puts the total overall death toll of lives lost off the coast of Newfoundland during that storm between 34 and 49.[265]

Michael Bruce, a survivor of that storm, remembered that the barometer showed clear weather until a few hours before the hurricane. On his father's boat, the *James and Martha*, at the time, he noted that the anchor buoy showed the first warning sign. It was bobbing down under the water and then up again, indicating that a strong tide was running—a warning sign of a storm. The captain—his father—ordered all the men out in dories to come back on board, and they headed for safety.[266]

When Captain John Spurvey of the schooner *Eleanor* arrived home at Aquaforte, he told the residents that he had seen a 70-tonne vessel off the coast of Ferryland drifting with bare poles, a piece of its jib in ribbons, and no foresail. Spurvey had fought hard to get home through heavy seas and wind, steering with only a cable, as he had lost his rudder in the storm.

St. John's newspapers reported extensive damage on the road from Topsail to Whitbourne. Trees were uprooted, fences blown down, and many roofs damaged. At Kelligrews, telephone and telegraph poles were cracked off or blown down. In St. John's, the storm was so severe that people got up in the night and dressed for fear that their houses would blow down. It was the worst storm in the memory of the oldest inhabitant in St. Mary's. The *Evening Telegram* referred to the storm surge this way: "A tremendous sea hove in and washed away large quantities of fish stored on the beaches."[267] According to the *Daily News*, St. John's sustained considerable damage of a minor nature, with a loss of electricity. The streets were littered with slates from the roofs of Water Street buildings and trees in Bannerman Park were uprooted. On LeMarchant Road, a store occupied by M.J. O'Brien Ltd. lost its roofing felt; inside, water rose to a depth of almost 1 metre. At Quidi Vidi, Andrew Snow lost his fish store, fish flake, and a stage, a loss of about $1,000. Also in Quidi Vidi, Robert Gulliver's house "crashed to the ground" at the height of the storm while he was sleeping. Holyrood residents lost their power and many of their gardens were destroyed.[268]

WEATHER SCIENCE

Recent hurricanes that have affected Newfoundland and Labrador include Gert on September 11, 1999, which moved a concrete breakwater over 15 metres[269]; Gabrielle, which blew over Newfoundland as an extratropical cyclone on September 19, 2001; Igor, on September 21, 2010; and Leslie, on September 11, 2012.

Often starting as weak disturbances off the coast of Africa, many Atlantic hurricanes migrate westward to the Caribbean, where they circulate clockwise around the Bermuda High (a sub-tropical high pressure system) and then move northeast along the Eastern Seaboard. Although most of these systems get caught in the prevailing westerlies and shift eastward and away from Newfoundland as they move north, a few follow a more northerly track that carries them over the island.[270]

1935

From wind and water to fire and earth: even if most forest fires start by human intervention, hot dry weather and wind add to their destructive potential. In August 1935, fires consumed almost 15,540 hectares of woodlands in the forests of central Newfoundland. The devastation covered a total area of 25,900 hectares but, because of the many rivers and ponds, the woodland area damaged was less. The fire started August 13 at the headwaters of Stony Brook, about 16 kilometres south of Grand Falls. Conditions were optimal for a large fire. The area had experienced an unusual summer with lower than average precipitation (the number of rainy days in June and July was the lowest on record for the area) and relative humidity, and higher than average temperatures. Wind velocities were above normal during June, July, and early August, and the forest floor was extremely dry. The fire that started on August 13 burned unchecked. Attempts to bring it under control started quickly but, even with the work of 1,200 men, the fire could not be contained. Lakes and streams served as barriers to the conflagration, and in some cases saved the lives of firefighters, who, cut off by the fire, had to wade into the water to survive. They escaped with only smoke blindness and minor burns.

On August 20, the fire still burned 8 kilometres from Grand Falls. But for a fortuitous change in the wind direction that day, the town would have faced certain doom. The fire continued until a low pressure system brought a downpour on the night of August 22. Another low followed, and by August 25, 13 centimetres of rain had fallen, successfully dousing the fire. These two wet weather systems were part of the first tropical storm of the season to come up from the Caribbean. Their presence was a stroke of luck for the residents and lumber companies of central Newfoundland—but not so for Grand Banks fishermen. The system that brought the rain damaged the fishing fleets; more than 50 men lost their lives in the storm on the Grand Banks and as far north as Labrador.[271]

1941

Torrential rains struck the Burin Peninsula and St. John's on August 2, 1941, with a total rainfall of 98.3 millimetres in 66 hours. Two employees of the United Towns Electric Company, Fred Connors and John Badcock, drowned when a dam at Little St. Lawrence broke.[272]

1976

In August 1976 the temperature in Botwood soared to a record high of 36.7°C.[273]

2002

On August 12 and 13, 2002, one storm in central Newfoundland produced over 10,000 lightning strikes and destroyed approximately 300 transformers, costing Newfoundland and Labrador Power about $1 million.[274] A large mass of moist unstable air mixed with a mass of warm air and resulted in an unusual amount of electrical discharges.[275]

TRADITIONAL WEATHER LORE

If you can hang a powder horn on the moon when it is new, it is a sign of stormy weather for that month.[276]

Hark! and then, above the rumbling
       in the chimney,
and the fast pattering on the glass,
       was heard a wailing, rushing
sound, which shook the walls as though
       a giant's hand were on them;
then a hoarse roar as if the sea had
       risen; then such a whirl and tumult
that the air seemed mad; and then,
       with a lengthened howl, the waves
of wind swept on, and left a moment's
       interval of rest ...

CHARLES DICKENS, *BARNABY RUDGE*

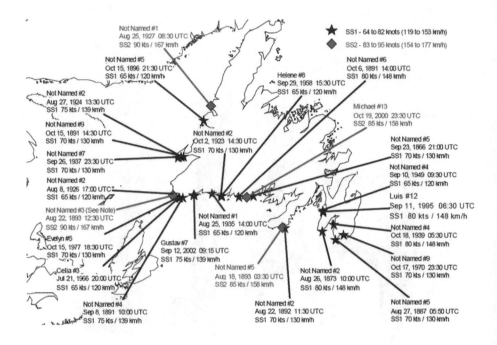

Not Named #1
Aug 25, 1927 08:30 UTC
SS2 90 kts / 167 km/h

Not Named #5
Oct 15, 1896 21:30 UTC
SS1 65 kts / 120 km/h

Helene #8
Sep 29, 1958 15:30 UTC
SS1 65 kts / 120 km/h

★ SS1 - 64 to 82 knots (119 to 153 km/h)
◆ SS2 - 83 to 95 knots (154 to 177 km/h)

Not Named #6
Oct 6, 1891 14:00 UTC
SS1 80 kts / 148 km/h

Not Named #2
Aug 27, 1924 13:30 UTC
SS1 75 kts / 139 km/h

Not Named #9
Oct 15, 1891 14:30 UTC
SS1 70 kts / 130 km/h

Not Named #7
Sep 26, 1937 23:30 UTC
SS1 70 kts / 130 km/h

Not Named #2
Aug 8, 1925 17:00 UTC
SS1 65 kts / 120 km/h

Not Named #3 (See Note)
Aug 22, 1893 12:30 UTC
SS2 90 kts / 167 km/h

Evelyn #5
Oct 15, 1977 18:30 UTC
SS1 70 kts / 130 km/h

Celia #3
Jul 21, 1966 20:00 UTC
SS1 65 kts / 120 km/h

Not Named #2
Oct 2, 1923 14:30 UTC
SS1 70 kts / 130 km/h

Michael #13
Oct 19, 2000 23:30 UTC
SS2 85 kts / 158 km/h

Not Named #5
Sep 23, 1866 21:00 UTC
SS1 70 kts / 130 km/h

Not Named #4
Sep 10, 1949 09:30 UTC
SS1 65 kts / 120 km/h

Luis #12
Sep 11, 1995 06:30 UTC
SS1 80 kts / 148 km/h

Not Named #4
Oct 18, 1939 05:30 UTC
SS1 80 kts / 148 km/h

Not Named #9
Oct 17, 1970 23:30 UTC
SS1 70 kts / 130 km/h

Not Named #1
Aug 25, 1935 14:00 UTC
SS1 65 kts / 120 km/h

Gustav #7
Sep 12, 2002 09:15 UTC
SS1 75 kts / 139 km/h

Not Named #5
Aug 18, 1893 03:30 UTC
SS2 85 kts / 158 km/h

Not Named #2
Aug 26, 1873 10:00 UTC
SS1 80 kts / 148 km/h

Not Named #2
Aug 22, 1892 11:30 UTC
SS1 70 kts / 130 km/h

Not Named #5
Aug 27, 1887 05:50 UTC
SS1 70 kts / 130 km/h

Not Named #4
Sep 8, 1891 10:00 UTC
SS1 75 kts / 139 km/h

HURRICANES THAT HAVE MADE LANDFALL IN NEWFOUNDLAND (TO 2002)

Environment Canada

SEPTEMBER

When you kill a spider, be prepared for a rainstorm within 24 hours.

1613

Crout records this detail in his weather diary for September 2, 1613:

> The winde at southeast verie fowlle weather and much raine
> all the day the winde beinge cold: the sune not showinge hir
> selffe all that d[aie] wher we wear forced to staie under a
> tree all this day and night conteinuinge rayninge untill the
> morninge: which by no mayner we coud not kept our bread
> drie but was all waitte.[277]

1770

The first temperature documented in Canada was recorded in September 1770 near Battle Harbour, Labrador. The mercury showed the equivalent of 29°C.[278]

When cats are very playful, they are said to "gale up the weather."

When my elderly aunt tried to get us children to calm down, she would yell: "Stop your galin'." Changes in humidity and air pressure can affect people and animals.[279] Perhaps this is why cats go galin'.

1775

In September 1775, a storm of uncommon force—the first recorded hurricane in Atlantic Canada and the deadliest to hit Canada—struck the coast and surrounding waters of Newfoundland and Labrador. Ranked as one of the 10 worst North Atlantic hurricanes in history,[280] it prompted this report in the *Pennsylvania Magazine*:

> A Person lately from Halifax to Cape Cod reports that he saw at Halifax, a particular account of the loss of several harbours of Newfoundland, in a violent storm on the 9th of September, amounting in the whole to more than four thousand men. It was said at Halifax to be computed that the loss of ships, fish, oil, and merchandize of various kinds, amounted to 140,000 L. sterling.

> An account received from Boston confirms the foregoing, and mentions—that nearly all the shallops employed that fishery, as well other vessels, were wholly lost; and those that rode out the gale, were chiefly dismasted, and otherwise much damaged, that many houses &c. were blown down, and that it would take the chief part of the spring to repair the flakes, they having received almost incredible damage.[281]

According to the Reverend John Jones of the Congregational
Church in St. John's, it was the "heaviest storm ever known in
Newfoundland. [The waves] swamped a considerable number of
boats and their crews."[282] It pummelled land and sea with such
ferocity that the governor, Robert Duff, reported the events
in a letter to the Earl of Dartmouth, Secretary of State for the
American Department, dated November 14, 1775:

> I am sorry to inform your Lordship that the Fisheries as
> well as the Trade of the Island of Newfoundland in the
> Month of September last, received a very severe stroke
> from the violence of a Storm of Wind, which almost swept
> everything before it. A considerable number of Boats, with
> their Crews, have been totally lost, several vessels wrecked
> on the Shores, and a number of those lying in the Harbours
> were forced from their Anchors and sustained much damage.
> The Fishing works in those places mostly exposed, were
> in a great measure defaced ...

Duff described the height of the seas as "scarcely ever known
before." According to Duff, two British schooners sank—one
stationed on the Banks, and the other on the northeast coast—but
only two people belonging to the crews of those vessels drowned.
Duff estimated overall damages to be not less than £30,000,[283] the
equivalent of about $6 million in 2014 Canadian dollars.[284] These
losses occurred in a wide geographical area and included Bonavista
and Trinity Bays, Fogo, La Scie, Conception Bay (including
Northern Bay, Harbour Grace, Carbonear, and Port de Grave),

Placentia Bay, Lamaline, St. John's and the Avalon Peninsula, and the islands of St. Pierre and Miquelon.[285] Judge Prowse said that St. Pierre and Miquelon, the northeast coast of Newfoundland, and west of the Burin Peninsula were the hardest hit.[286]

In the *Annual Register* for 1775, it was reported that "at St. John's and other places there arose a tempest of a most particular kind— the sea rose 30 feet ... 700 hundred boats and 11 ships were lost with most of their crews." For days after the storm the fishermen would draw their nets ashore and find 20 to 30 bodies in them each time.[287]

Duff also included a description of losses suffered by the French fishery. St. Pierre and Miquelon lost 400 men, a monumental blow to the tiny islands: "Of the 40 French fishing boats who were then fishing on the Grand Banks, the 22 that belonged to the inhabitants of the islands were lost, along with 50 small boats and the destruction of a number of stages." Over 9,400 quintals of fish and 30 tons of oil were lost and nine banking vessels and 20 boats from Miquelon sank.[288]

Reverend Lewis Anspach, a Church of England clergyman, schoolmaster, and circuit court judge in Newfoundland from 1799 to 1812, described the storm:

> On the 12th of September, in the year 1775, this coast was visited by a most terrible gale of wind. In Harbour-Grace and

Carbonear all the vessels in the harbour were driven from
their anchors; but the inhabitants of the North Shore of
Conception Bay suffered with greater severity. They, even now,
with evident signs of dread and horror, show a cove where
upwards of 200 fishing vessels perished with all their crews.[289]

At Placentia, many fishermen lost their lives, their stages and
flakes were damaged and their boats sunk, and flood waters
rose 1 metre, forcing residents to take refuge in the rafters of
their homes.[290]

Over 40 fishing vessels from the British Isles were in Northern
Bay and may have been the first to see warning signs of the
storm. After three weeks of calm, idle fishermen began to fume
over the lack of squid. According to Leo English, they cursed the
elements, the fish, and everything else, "profanely and defiantly,
they besought their Maker to send wind. Let it come, no matter
how. Send it so that a man could not hold the blade of a knife to
windward." English continues:

> Darkness and dead ominous calm brooded over sea and land.
> The squid came that late summer afternoon and so abundant
> and omnivorous that the oldest seamen grew terrified as they
> filled the ships' boats. Over the horizon to the southeast there
> spread an orange hued glow. Then wisps of wind, that slowly
> gathered strength and increased to gale force culminated in
> the fierce violence of hurricane.[291]

English reported that one small schooner escaped the storm. Its captain had the foresight to recognize the signs of a storm and had sheltered in nearby Ochre Pit Cove. In Northern Bay, there was no shelter: over 300 people died when their ships sank. Their mass grave on shore is marked with a cross, still visible in 2014.[292]

Legend has it that only one boy, Billy Gill, survived that dreadful night. Some sources contend he was in the cuddy of the boat,[293] but English insists that the boy was lashed to the helm and made it to safety when a huge wave lifted the boat out of the water, carried it up onto the sand, and wedged it between large trees. Bones washed ashore for years after the event. Residents say the sands of Northern Bay are haunted, and some say they've heard ghosts, or "Hollies," call out a warning when a storm approaches. "[T]he bawling of the Hollies sounds just like a crowd of frogs," says Tom Mullaly of Northern Bay. "Sometimes they are up to mischief, but at other times the Hollies can be helpful." John Hogan, a fisherman who died in 1972, saw the Hollies just before a storm: they looked like regular men and helped him haul his fishing gear and load his crishlow [basket] with kelp.[294]

Labrador, too, felt the wrath of a storm on these same dates. Explorer George Cartwright described the early storm signs in his diary on Sunday, September 10, 1775. On the coast, the day was dark, thick with fog in the evening, and the night rainy. His Monday September 11, 1775, entry notes:

At one this morning it began to blow hard; at five the gale was heavy, and in half an hour after, our cable parted: we were then near driving on shore upon Western Point, which is shoal, and rocky; but we got her before the wind, ran up the harbour, and let go the other anchor between a small woody island and Earl Island, where we brought up in four fathoms and a half of water, over a bottom of tough black mud, and there rode out the gale. The water was perfectly smooth, but the wind blew so excessively hard, that the vessel was frequently laid almost on her beam ends, the tide making her ride athwart the wind, and the spoondrift flew entirely over her ... I went on shore and was informed, that the tide yesterday, flowed two feet higher than usual; that it rose two inches high in the house; and that the violence of the wind was so great, as to turn the bottom up of a sealing skiff, which lay on Rocky Point. I never experienced so hard a gale before.[295]

Marine geologist and researcher Dr. Alan Ruffman says that the storm of 1775 became a "trapped fetch" with two types of wind: one moving in the circular motion, the other propelling the storm farther across the surface of the earth. He describes the consequences: "When you're on the right-hand side of the storm the circular velocity and the forward velocity add together and that's what gives you the very serious winds and the very large storm surge."[296] Storms that come up the coast to Newfoundland weaken as they pass over cooler waters, but their structure can change as they move toward the province. The size of the wind field usually grows, particularly to the right of the storm track, as a storm moves toward the island. A storm's forward motion often increases in speed as it heads into higher latitudes.

Ed Rappaport, a hurricane specialist at the National Hurricane Center in Miami, speculates that the 1775 storm was probably stronger than 2010's Hurricane Igor; its centre likely passed 100 to 200 kilometres closer to the island of Newfoundland. Igor did make landfall but tracked southeast of Cape Race before moving northeast of the island.[297] Wave heights of 6 to 9 metres reported during the 1775 hurricane were caused by storm surge and waves, not a tsunami. This wave height would have needed the substantial wind speed usually associated with intense hurricanes. By checking Lloyd's list for September 1775 and shipwreck data for North America for the same period, Rappaport and his colleague, José Fernández-Partagás, estimated the storm's death toll to be 4,000.[298]

Several conflicting stories exist about the hurricane that hit Newfoundland and Labrador between September 9 and 12, 1775. Some older sources refer to this storm as the Independence Hurricane. Of the large numbers of people killed by the hurricane off the coast of Newfoundland, many were seamen from Britain and Ireland. The loss of British fishermen available for naval service may have weakened the British response to the American Revolution, and this might be part of the reason why it was known as the "Independence Hurricane." An earlier storm hit North Carolina and Virginia on September 2 to 6 coinciding with the opening manoeuvres for the War of Independence. This was probably the real "independence" storm.[299] At one point it was believed that these two events were one long-lasting storm, stretching from September 2 to September 12. Meteorologists no longer believe this. The Independence Storm was the worst and most destructive hurricane ever to hit North Carolina. Another hurricane struck Nova Scotia on September 9. These hurricanes were two separate, catastrophic events.

As serious as it was, the Independence Storm of September 2 to 6 was not as lethal as the storm that hit Nova Scotia and Newfoundland between September 9 and 12 that same year. The North Carolina storm claimed hundreds of lives, drove some ships ashore, and caused others to be lost at sea. The storm possibly made landfall close to Cape Hatteras and continued up the coast as far as New England and Massachusetts, causing further destruction before moving out to sea. Classed by Rappaport and Fernández-Partagás as the 114th deadliest tropical cyclone in the North Atlantic, the North Carolina storm caused more than 163 deaths. The Newfoundland storm of September 12, however, ranks seventh on the list of deadliest tropical cyclones.[300]

WEATHER SCIENCE

Low pressure systems can cause a pressure surge in which water levels rise slightly—but the winds associated with a low pressure system have a greater impact. A strong low pressure system can add a wind-driven surge to the pressure surge, as winds circling

**THESE TOWERING CUMULUS, MOST OFTEN SEEN IN
NEWFOUNDLAND IN THE SUMMER, FORETELL HEAVY RAIN
SHOWERS OR THUNDER SHOWERS**

Environment Canada

---

the system pile water up near the storm's centre.[301]

1837–38

In the *Bulletin Scientifique* published in 1849, Moravian missionary
C.B. Henn described the weather in Okak on the coast of Labrador.
In August 1837, he noted, the snow had not yet started. September,
however, brought heavy night frosts and, by September 10, the
mission experienced the first snow and ice. This did not stop
the missionaries from harvesting 3,500 potatoes. Snow and frost
began in October, and Henn recorded temperatures of -6°C to
-10°C for this month.

In November, the snowstorms came and temperatures hovered between -6°C and -14°C. December and January were mostly beautiful and clear: temperatures ranged from -19°C to -34°C, with little snow. The bay froze over, and the Inuit went out on the ice for several hours at a time to look for seals. Temperatures rose in February and March, with lows of -13°C to -15°C. Toward the end of the month, temperatures dropped to -29°C. By the end of April, 4 to 5 metres of snow covered potato gardens. The missionaries covered the snow with ashes and other black materials to help the sun melt it. This helped them locate the tops of their 2-metre fence posts. Henn engaged the help of 30 Inuit to dig the snow out of the garden so the beds could be prepared for planting. On May 3, they planted beets in a patch of garden not far from a 2- to 3-metre-high wall of snow. It took until the end of May, with temperatures hovering around 0°C, for the snow to melt enough so that the Moravians could plant potatoes, yellow and white turnips, beets, parsley, and onions. Toward the end of June they could plant the rest of their vegetables in the snow-free parts of the garden, as on the western part of the garden—where there had been 6 metres of snow, only 1 metre remained.

June and early July brought beautiful weather with daytime temperatures between 28°C and 31°C, and open water on the bay. On July 4 the frost returned and the bay filled with slob ice. The Moravians had covered their crops with straw to protect them against any sudden temperature change. For the rest of the summer temperatures yo-yoed from the warm temperatures normal for the season to zero degrees combined with thunder and lightning storms and "ice-cold sea fog."[302]

1846

On September 19, a few months after the Great Fire of 1846, which destroyed three-quarters of St. John's, a storm struck the city, plunging residents into an even more desperate situation:

> The most severe storm of wind and rain ever witnessed commenced on Saturday last. The weather for some days had been of a nature to lead us to expect a violent change and the short "chafing of the elements" coming and going in fitful evolutions, evidently portended that they were but "the beginnings of the end." On Thursday there had been a fresh breeze almost amounting to a gale, from N.W. but Friday broke upon us softly and moderately, and closed with nothing more than a light rain save a sickly and murky appearance in the East which tallied but too well with the previous indications of an equinoctial blast and the still falling "glass." On Saturday morning, early, a strong breeze from E.S.E.; and this continued to increase through the former part of the day, until it eventuated in what we have been accustomed to call a storm. By about 3 o'clock in the afternoon the wind veered rather to the Northward of East, and blew a perfect hurricane; the rain was poured upon the earth as though all the windows of heaven were opening; and people seemed, looking forth from their crowded and (comparatively with what they had been accustomed to before the fire) uncomfortable homes,

as if in anticipation of some fresh calamity. We deeply regret that such anticipation was awfully realised, not only in the destruction of property to a very considerable amount, but in a melancholy loss of human life.[303]

It took just hours, from 6:30 a.m. to 8:30 a.m., for the barometer to drop from 1009.14 to 998.98 millibars. With the wind blowing from east-northeast, a "perfect hurricane" developed by 2:30 p.m., about which time the wind veered to north-northeast. The gale continued unabated all afternoon, with the wind and rain increasing until about 4:30 p.m., when the barometer began to rise again. Powerful gusts of wind and torrential rains continued. About 10 p.m., the rain stopped, and by midnight the gale had moderated. Late in the afternoon (some reports say at 4 p.m., others 5 p.m.), the Native hall, a building under construction on the site of what would become Bannerman Park, collapsed. It sheltered families that had been left homeless in the Great Fire three months previous. The *Times* reported that when the collapse occurred "the shrieks of the sufferers and of their friends were enough to make ... the heart bleed." A young brother and sister by the last name of Duggan, 5 and 10 years old, were killed in the collapse. Their mother escaped with her life but was badly injured. Rescuers found her with "an emmense weight of timber pressing her to the earth."[304]

St. Thomas' Anglican Church, known as "The Old Garrison" because it served the soldiers stationed in nineteenth-century St. John's was "removed four inches from its former position."[305]

The rivers in St. John's overflowed their banks, rising in some places to 3 metres above normal levels, causing floods of raging water full of storm debris. Much of the flooding resulted from the state of the land:

> [By the] prevalence of extreme drought throughout the summer [the land] had been converted literally to powder, and thus rendered almost impervious to the water. The rain descending in such torrents was thus forced to form its own channels to the nearest streams, which became soon swollen to such an extent as to occasion great damage to the numerous bridges.[306]

Bridges across Rennie's Mill River were washed out from Long Pond to Quidi Vidi; even the stone King's Bridge was washed away. Many of the boats attempting to seek shelter in St. John's harbour swamped in the Narrows with all on board lost.

At Quidi Vidi, the waves swept away oil, nets, boats, fish, flakes—everything in reach of the seething vortex. The estimated damage was "not less than £1000 ... the greater part of which falls on the poor fishermen, the proceeds of whose summer's labours were ready for market and were destroyed in a few hours."[307] High winds and heavy rains wrecked and dismasted vessels in outports and harbours along the east coast of the island, swamping them in high waves or pulling them from their moorings to cast them up on serrated rocks. Fishing stages and flakes were swept away; in many cases, a whole season's catch was engulfed in the spume.

Logy Bay, Middle Cove, Torbay, Flatrock, and Pouch Cove
suffered destruction and loss. A vessel belonging to Mr. Coyell
of St. John's, fully loaded with fish and oil and bound for
Twillingate, drove ashore in Freshwater Bay, and disintegrated.
All escaped drowning except one. The *Lavinia* from Harbour
Grace, bound for the Naples with a load of fish, sank at Pouch
Cove with one man drowned. The Carbonear and Harbour Grace
packet boats rode out the gale by cutting away their masts, but
at Portugal Cove the Brigus packet boat sank. Forty boats were
swamped or smashed against the rocks in Bay Bulls:

> Each succeeding day furnishes some new and distressing
> intelligence of disaster by the fearful and terrific storm of
> the 19th inst. Amongst the losses of life on the occasion we
> sincerely regret to have to record the affliction which befell
> the family of Mr. Thomas Brine of Bay Bulls, whose two
> sons, both fine young men, were crushed to death by the
> falling of a large store.

Thomas Brine's life was spared, but both limbs were fractured.

More grave news came in from Renews. Twenty boats were lost
in the town, and seven men drowned, including John Butler and
his two sons from Port de Grave. Magistrate Thomas Hutchings
and Episcopal clergyman John Roberts from Bay de Verde wrote
a letter to the editor of the *Newfoundlander* telling of the fate
of Grates Cove, where the wind destroyed fishing stages and
flakes as well as 60 out of 70 fishing skiffs lying at anchor in the

cove. Catches of fish and oil were ruined, leaving the fishermen with no means of survival for the coming winter. The letter begs "Her Majesty's Government in Newfoundland [to] be graciously pleased to consider the emergency of their case, and provide some relief, as was done in the case of those who were distressed by the late fire." It listed other communities that were hit particularly hard: Red Head Cove, Old Perlican, Lance Cove, Hant's Harbour, New Perlican, Heart's Content, Low Point, Island Cove [now Lower Island Cove], and others.[308]

The *Patriot* reported that the storm was one of the most destructive storms of wind and rain ever in the area:

> We lament to record, as the effect of this visitation, (so far as already ascertained) a great loss of life and property ... a vast number of fishing stages and flakes in various harbours on the eastern coast and in Conception Bay [destroyed] ... Houses have been blown from their foundations in two lamentable instances crushing beneath the ruins, some of their unfortunate tenants. Trees in almost every direction have been uprooted from their beds or broken in pieces by the fury of the gale.[309]

At Hant's Harbour, three schooners—*Swallow*, *Daniel O'Connell*, and *Curlew*—sank. Henry Parsons of Freshwater lost his schooner south of Carbonear and £400 worth of property belonging to the Messrs. Collings of Spaniard's Bay.[310] The storm also pummelled the southwest coast. In St. Mary's, groups of boats laden with the fish harvest were ready to sail for St. John's. Every craft among them, except two skiffs, met their doom.

At St. Pierre and Miquelon, vessels were driven on shore and destroyed. The *Comet*, holding passengers including the assistant judge and officers of the Southern Circuit Court, sank, as did boats from Nova Scotia, Prince Edward Island, and France. A *Royal Gazette* reporter wrote on September 29:

> Previously to the occurrence of this destructive hurricane, the situation and prospects of the Colony, from a succession of indifferent fisheries and the destruction of its capital by fire, were sufficiently gloomy—they are now alarmingly so.[311]

This storm, however, primarily affected "a class of persons very ill able to sustain it ... the outport fishermen, many of whom have lost the means of support they possess and are left at the approach of a long and dreary winter, destitute of the necessities of life with no visible means of obtaining them."[312]

Repercussions from this storm were felt as far away as Marblehead, Massachusetts, where the families of fishermen killed while fishing on the Grand Banks grieved for their losses. One man who survived the storm on the Banks recounted:

> One moment we were on the very top of the highest wave, and the next would be dashed down apparently hundreds of feet ... as if one were suddenly thrown from a precipice into

a pit of seething, angry waters ... Nothing could describe
the agonizing fear that beset us all in the hours in which we
were compelled to sit idly in that cabin, silently waiting what
seemed inevitable destruction.[313]

Eleven vessels and 65 men and boys from Marblehead were lost on
the Grand Banks during that year's storm. This was the beginning
of the decline of the fishery in those parts.

1907

Since the early days of colonization, large storms continue to blast
into Newfoundland. Unlike modern hurricanes with names like Igor
or Leslie, the storm that hit in September 1907 was known simply
as "the storm of 1907" or "the Twillingate storm." Its hurricane-
force winds hit Twillingate and the Fogo communities of Joe Batt's
Arm, Barr'd Islands, and Little Fogo Islands. High winds and raging
seas crushed and swamped fishing boats, flattened stages and
stores, and tore down fishing flakes; the storm's effects were felt
from Labrador down the northern and eastern coasts of the island
to Trepassey. Telegraph lines and government lines were down in
the west and north, and on the following day, September 20, the
Reid Railway Company could not contact any stations west of Terra
Nova.[314] About 32 kilometres of the main government telegraph
line between Port Blandford and Gambo were completely wrecked,
poles uprooted, and wires buried under fallen timber.[315]

The damage in Twillingate was unprecedented. Thirty schooners
were tossed upon the rocks like toy boats. The *Daily News* reported

**SCHOONER WRECKAGE AT TWILLINGATE AFTER THE STORM OF
SEPTEMBER 17, 1907**

Captain Harry Stone Collection, MHA

that of the many schooners anchored at Twillingate only two
escaped from being driven ashore. One by one, the ships broke
anchor and drove up on the beach—one ship's mast went through
a fishing store window. Much of the valuable summer fish cargo
survived intact, and some vessels sustained only minor damage.
Only 10 of the schooners were insured. Along the waterfront,
flakes, stores, and wharves—including the government wharf—
were destroyed by the storm.[316]

High winds and seas drove the schooner *Nina Pearl*, skippered
by Captain Ambrose Payne, and the *Osprey*, owned by Mr. Earle,
ashore, but their crews successfully re-floated both vessels. The

schooner *Leslie E.*, captained by William Snow and bound for
Flower's Cove, went down in the storm, but the captain and crew
escaped. Reports came in from the Straits and Blanc Sablon that a
heavy gale was raging but no damage was done. The schooner *Poppy*,
en route from St. John's to Dog Bay, rode out the storm:

> At 10 o'clock [p.m.] a perfect hurricane was raging. The seas
> went over her fore, and aft, and everything movable on deck
> was washed overboard. The heavy seas broke in both bulwarks,
> washing water casks and other loose articles overboard,
> carrying away her foresail which was rent in pieces and the
> topmast fell and got caught in the rigging.[317]

Farther down the shore at Bonavista, the carnage was replayed.
Four fishing schooners—*Harold F.*, *Olive Branch*, *Planet*, and
Reliance—as well as numerous small skiffs and other fishing boats,
lost their anchors, and went ashore. The Bonavista lighthouse
keeper sent this message: "Light out; Long bridge and north side
gone; wood and glass; impossible to land."[318]

A Norwegian vessel, the *Snorre*, chartered by the Bonavista firm
of James Ryan, foundered just after she had completed her maiden
voyage across the Atlantic. All except five of the crew were onshore
when the storm hit. The men of Bonavista valiantly rescued
three of the five left on board. For this they were awarded
medals of bravery from King Haarkon of Norway. The other two
crew members, both boys, were lost to the sea.[319]

At Trinity, the *Effie M.* went down with 16 to 18 on board. All were lost. Only 11 bodies were recovered when the men from Old Perlican and the surrounding area went out the day after the storm in punts and skiffs to search for the dead.[320] At Harbour Grace, fences, stables, and barns blew down: "So great was the force of wind that trees which stood the gales for the past 60–70 years were uprooted—as one would scarcely think possible."[321] A schooner coming from St. John's, the *Hettie*, captained by Joseph Morris, limped into the harbour that day, a total wreck. She had lost her cargo of freight, and one Harbour Grace grocer had reportedly lost $1,000 worth of goods on the schooner.[322]

The *Evening Telegram* reported that at Cape Spear the seas were so heavy that the spray reached the top of the lighthouse.[323] Newspaper headlines included: "From Grates Cove Thirty Punts and All Stages Swept Away," "Tragedies of the Storm! A Tale of Death and Disaster," and "Man Dashed to Death at Arnold's Cove." A *Daily News* writer described the havoc wrecked by the storm on Newfoundland:

> Surely the price of admiralty has been paid by British blood and yet the Moloch of the deep remains unsatisfied. Newfoundland's contributions to the sacrifice of blood and tears have been recorded in the homes and hearts of succeeding generations, but it seems as though only part of the price has yet to be paid and we must feed our seas for a thousand years for that is our doom and pride. The storms

GEORGE'S BANK COD FISHERY
NOAA

---

of Wednesday and Thursday will rank with those of 1846 and 1885 in the history of our island home. Day after day brings its tale of woe.[324]

1916

An *Evening Telegram* writer noted on September 26 that it had been 26 years since a hurricane equal to that of the 1916 hurricane had swept over the island. Not only was it fierce but it was also "felt in its intensity in every section of the coast and in every harbor and bay on land and sea."[325] Under the headline "City Storm Swept!" was this description: "In the Throes of a Fierce

Hurricane ... towns and suburbs the scene of much wreckage—
anxiety felt for shipping." Winds blew up to 100 miles an hour,
lightning flashed, and everything shivered "as if from the effects
of a miniature earthquake."[326] No one heard thunder; perhaps it
was drowned out by the shrieking winds. Lightning knocked out
power in St. John's, and windows were blown out of thousands
of residences, churches, and public buildings. One house on Lime
Street capsized, and residents barely escaped with their lives.

In St. John's, Parade Rink (also known as Prince's Rink), near
the site of the Newfoundland Hotel, had its roof blown off and
sides smashed in. Both Gower Street United Church and the
courthouse lost their clock faces and several slate roofing tiles.
Methodist College Hall lost roof slates and had windows blown
in. The winds came from the southwest by south, then veered
around to the northwest at about 3 a.m. on the 17th. Many people
stayed up all night in fear of the storm's power. Worried about
the possible collapse of their houses, they worked hard to secure
their properties. Trees and fences blew in all directions, and
even several chimneys were knocked down. Falling trees broke
electrical wires and plunged the town into darkness.

High winds took off the entire roof of the chapel of Our Lady
of Good Counsel at Cathedral Square and damaged the roof of
St. Andrew's Presbyterian Church. The stained glass window in
Cochrane Street Methodist Church was smashed. St. Joseph's
Presbytery on Signal Hill shifted off its foundations, and the tower

on the Bank of Montreal building twisted with the force of the winds. Farmers and milk vendors coming into town reported that the roads were littered with chimney tops, barn roofs, outhouses, and fences. Roofs were blown off the barns belonging to J. Brennan and Joe Butler, and Edward Skeans of Topsail Road lost both the roof of his barn and his house. On Leslie Street, Mr. Parsons's roof blew 91 metres into the middle of the street, and his house lost one of its sides. Considering the solid construction of many of downtown houses (many supported by huge wharf pylons), the storm damage is frightening. Boats at the Battery were "smashed to atoms" on the rocks, and stages were destroyed.[327]

At Logy Bay, the storm destroyed a seal oil factory owned and operated by Phillip Malone, William Weir, and A. Snow. Pushed by the extreme force of the gale, the factory, which was situated on the north side of the bay, toppled over into a small stream below it. The factory's contents included two months' stock of cod liver oil, boilers, cooling tanks, and other equipment, all of which were lost. In Logy Bay, the wind blew Carrigan's stage, including a large quantity of cod, 91 metres from its foundation.[328]

Patrick Manning from Torbay, who was in the woods on Sunday evening looking for his horse, got lost in the fog and was forced to remain there all night in the storm, where he received a "drubbing." He returned home safely on Monday morning.[329]

In the Goulds, just outside St. John's, two homes lost their roofs, and a barn full of hay collapsed. The September 26 *Evening Telegram*

THE EAST END OF ST. JOHN'S FROM THE ROOF OF THE NEWFOUNDLAND HOTEL LOOKING TOWARD SIGNAL HILL, INCLUDING PRINCE'S RINK, CA. 1925

ASCD

---

mentioned the carpenters, masons, and other workmen repairing roofs, buildings, and other damage done by the gale in St. John's.

In Harbour Grace, barns were blown down, windows smashed in the shops on Water Street, and schooners, motorboats, and other small boats lost offshore. The schooner *Eclipse* drove ashore at Harbour Grace, and the *Dorothy* sank near the government wharf. Munn's wharf sustained damages when the barquentine *St. Simon* collided with it. The storm also wreaked considerable damage on Pouch Cove, but power lines were down, and no communication with the outside world was possible for some time. In Bay Roberts, two coal boats collided with the wharf, damaging both the boats and the wharf. At the same location, the roof of the Union Trading Store was carried away.[330]

Along the Southern Shore, schooners were driven ashore at Bay Bulls, Mobile, and Witless Bay; small craft and fishing stages were destroyed. Six houses were blown down in Trepassey, along with two fishing stores full of fish. A week later, news came in from Bay de Verde: the schooner *Mary Alberta* belonging to Moses Blundon, a young fisherman, sank with 280 quintals of fish, nearly all of his summer catch. He had no insurance. More than 40 other fishing boats and several motorboats from the community were also destroyed. Losses were estimated at between $8,000 and $10,000.[331]

On the high seas, Captain Kean on the *Portia* made a run for Trepassey when the winds reached cyclone speeds and the ocean began to seethe. The waves became "mountains high" and the night intensely dark, but occasionally Trepassey lit up with vivid flashes of lightning.[332] The *Evening Telegram* reported that "Grave fears were entertained" for the fishing vessel *Annie*, but it was hoped that she had found shelter from the storm in some small harbour.[333] The *Annie* was lost between Ragged Islands in Placentia Bay and St. John's with a full cargo of dry fish and a crew of five.[334] The master and crew of the schooner *Avis* from Quebec abandoned their ship and a full cargo of lumber 48 kilometres west of Miquelon. All people were rescued,

and returned to St. John's. The *Bonnie Lass*, owned by Michael
MacDonald of Newbridge, Salmonier, was not so lucky and was
found at Trepassey Bay with her hull partially submerged and
her bow up. The bowsprit and the foremast were intact, but
the double-reefed foresail on the foremast was shredded and
the mainmast unstepped. Because the schooner did not run
aground, the *Telegram* reporter postulated that it had either
swamped or capsized. No bodies were ever found.[335] The *Theresa
M. Grey*, sister ship of the ill-fated *Bonnie Lass*, arrived at North
Harbour "with not a feather out of her."[336]

The SS *Viking*, captained by Cyrus F. Taylor, arrived in St. John's
from North Sydney but had "got[ten] the full force of Sunday
night's storm. She was hove to for sixteen hours, during which
she was badly buffeted. The wind raged from the southwest and
heavy seas broke over her fore and aft inundating the decks."
The *Viking* was unharmed, but the trip from North Sydney to
Newfoundland, with a cargo of coal for Bowring Brothers, took
75 hours.[337]

The lighthouse keeper at Cape Race reported "heavy storm
blowing south and then turning round to Southwest. Terrific sea
on."[338] The next day he reported: "No vessels sighted today."[339]

The fishing industry suffered severely that season; many local
craft of all sizes either had been smashed or significantly
damaged.

WATERFORD RIVER FLOODING NEAR THE KILBRIDE BRIDGE, SEPTEMBER 15, 1948

City of St. John's Archives

1948

On September 14, 1948, heavy rains caused mudslides that collapsed three houses in the St. John's area.[340]

2010

Hurricane Igor struck the island of Newfoundland on September 21, 2010, with winds up to 140 kilometres per hour; peak winds of 172 kilometres per hour were recorded at Cape Pine. The storm washed out roads and turned others into rivers, several with their own waterfalls. Igor dismantled bridges, crumbled culverts, and knocked out power, forcing residents from their houses. The areas of highest rainfall and

wind were in eastern Newfoundland. In Bonavista, more than
200 millimetres of rain fell in 22 hours; St. Lawrence, on the
Burin Peninsula, received 239 millimetres in 20 hours. When
their roads and bridges were washed out, many communities
were essentially cut off from the outside world. Newfoundland
Power estimated that about 50,000 customers, households,
and businesses were without power on Tuesday, September 21.

Hurricanes generally weaken as their energy dissipates over
the cooler northern waters near Newfoundland. A stationary
front with a low pressure trough in the upper atmosphere,
however, fed energy back into Igor.

Food supplies ran frighteningly low, and Newfoundlanders
were faced with the fact that, even in 2010, the island is highly
dependent on the mainland North America for food and other
necessities. Twenty-two communities declared a state of
emergency and 150 were totally isolated, their roads washed
out and no electrical power.[341] It was not until Sunday,
September 26, that power was restored to most of the island,
and many schools did not re-open until Monday.

One death resulted from the hurricane. Eighty-year-old
Allen Duffett of Random Island was standing in a friend's
driveway when the pavement gave way underneath him and
he was washed away in the flood waters. His body was not
found until several days after the storm, buried under the
rubble on a beach.

HURRICANE IGOR OVER NEWFOUNDLAND

Environment Canada

The cost of Igor was enormous: $65 million in insurable claims
and over $120 million in non-insured claims.[342] The Army was
called in to help with cleanup. Four regiments of engineers
landed at Argentia and fanned out across the hardest hit areas.
They re-built bridges and helped out wherever possible. In all,
over 1,000 military personnel, along with heavy equipment,
Navy ships, and Sea King helicopters, assisted in this effort.[343]

One of the noteworthy features of Igor was its diameter;
the breadth of the area it affected was broader than that of
previous storms. Hurricane Igor was classed as a rare event in
the annals of Newfoundland weather. The storm left on the
morning of Wednesday, September 22.

It is interesting to note that the storm on February 15-16, 1959,
which caused the Battery Avalanche, had estimated wind speeds
of 217 kilometres per hour.[344] The Canadian Hurricane Centre,
however, called Igor "Newfoundland's most damaging hurricane
in 75 years."[345] Another hurricane of similar intensity had struck
the island in 1935, killing 49 seafaring Newfoundlanders.[346]

TRADITIONAL WEATHER LORE

When wild animals take on a thick coat of fur in autumn, it is a sign of a bad winter to come.

WEATHER SCIENCE

The jet stream is a narrow band of high-altitude, fast-moving air which forms a dividing line between warm and cold air masses. It constantly shifts its position and strength. If the jet stream lies to the south, it keeps the warm weather at bay, but when it hovers farther north, warmer temperatures prevail. Storms and highly variable conditions are more common in areas over which the jet stream passes, while regions farther away usually experience more stable weather.[347] The record snowfall of 2001 was due to the jet stream's position: it hovered just south of Newfoundland for much of the winter, resulting in many winter storms and sustained colder-than-usual temperatures. The lack of mild weather and rainfall allowed the snow to accumulate and remain on the ground all winter long.[348]

RATTLE BROOK BRIDGE, ON THE BURIN PENINSULA HIGHWAY, WAS SEVERELY DAMAGED BY HURRICANE IGOR, VIRTUALLY CUTTING OFF THE PENINSULA FROM THE REST OF NEWFOUNDLAND

Fire and Emergency Services, Government of Newfoundland and Labrador

NIMBOSTRATUS CLOUDS OVER GANDER, JANUARY 26, 2014

Photo by Rodney Barney, Environment Canada

OCTOBER

If rain takes the first snowfall away, it will be a stormy winter, but if the sun melts it, cold dry weather will come.

1612

From Crout's diary:

> October 12, 1612—in the morninge the wind at weste
> northweste all the fore noone verie close weather and much
> winde in the after noone verie faire sune shininge but verie
> much wind untill night the night verie calme but very colde
> we landed out of the French shipp left ther by captaine
> Eastone some 15 tonns of Salts upon the beache allso lefte
> 2 cabells and a anker which we lefte ther allso the some 60
> Fadome the shipp 120 tonne[.][349]

> October 17—in the morninge the wind at north weste this
> morninge we departed from Harbour de Grace abowte 7 of
> the clocke all this daie verie faire sune shyninge untill night
> and at night abowte the setting of the sune we ankered
> in Green Bay of the wester side the wind at northwest in
> the night we had some showers of snowe this night after
> midnight we wayed anker to go abowte Backaloo but ther
> finding the wind contrarie bonding too and froo.[350]

A PLENTIFUL CROP OF DOGBERRIES

Photo by Craig Purchase, Memorial University of Newfoundland

1796

Here is Thoresby's account of October weather written while
he was on board a ship from England as it approached the
Grand Banks:

> October 5, 1796—It blew a gale the last night, and doth at
> the present. We have seen an innumerable number of birds,
> which indicates that we are near the banks; and the sailors
> say, that one species of them is a sure token that there are
> vessels on the banks. At twelve o'clock this day we tried to
> sound and found the water forty fathoms; it gave all on
> board a degree of pleasure.

October 6, 1796—... A sloop of war went past us from St. John's, but we could not come near enough to speak with her, for it blew a tremendous storm, such a one as I never saw before, indeed it was with the greatest difficulty that we could hear each other speak on board our own vessel.[351]

TRADITIONAL WEATHER LORE

An abundance of dogberries on the trees are a sign of a hard winter to come. The higher up they are, the harder the winter will be.

1874

The October 10, 1874, entry in the Ryan diaries noted:

Change of moon this morning has brought change of weather. Wind stiffish westerly. Air humid. A little rain fell during the night.[352]

WEATHER SCIENCE

Nimbostratus clouds, grey or blue clouds that cover the whole sky, usually indicate that a storm is on the way.

1885

The Labrador gale of 1885 started on Sunday, October 11, and did not abate until the evening of October 12. What started as a southeast wind in the morning of October 11 had veered to northeast by noon, increasing in fury and bringing heavy snow flurries and bitter cold. Eli King, whose ship the *Release* went down in the tempest, described the beginning of the storm: "The wind rose quiet as the sunrise ... slowly increased to a gale, and at eight o'clock on Sunday night it blew at its height, a furious and restless hurricane."[353]

Twenty-five people lost their lives on the *Release*, nearly all of whom were women and children. Another schooner at White Bear Islands, the *Hope*, came away from her moorings and almost succeeded in reaching the shore before she broke up, taking the 13 left on board, mostly women and children, down to the deep with her.

Most of the storm's impact was felt at sea. Vessels, heavily loaded with the summer's catch, also carried many of the fishermen's wives and children, who had come along to help with the Labrador fishery. In addition to the men that drowned, many women and children died as the ships went down. Most of the island schooners had left the Labrador coast for the season, or the damage and loss of life would have been even greater.

Of the many stories to come out of the terrible storm, one about a man and his son fighting for their lives in the roiling surf is particularly poignant. The man had his son in his arms, but the son had a broken arm and could not help his father. When the boy realized they would die if he didn't help, he loosed his arm from the bandage and struck out, only to be taken by a great wave and carried onto the land, and safety. The report does not state if the father survived.[354]

A vessel from Tilton, near Bay Roberts, went aground on the rocks while attempting to outrun the storm, but managed to free itself and put ashore at Mark's Island. All the passengers survived, with the exception of two young girls, who died of "fright, exposure and weakness."[355] An unnamed woman from the north shore of Conception Bay found herself in the waters off White Bear Island with no means of getting ashore until two Newfoundland dogs which were on board her vessel came to her rescue. She grabbed their backs and was towed to land.[356]

Off the coast of Labrador, a young man by the name of Reardon made it ashore from a vessel that was breaking up but went right back to rescue a young woman who had called out to him for help. Just before he got to her, he witnessed the deaths of two other women when punts fell on them.[357] A captain of a ship caught in a storm like the Labrador gale of 1885—with winds reaching an estimated speed of 193 kilometres per hour[358]— likely prayed that its anchors would hold and, as a precaution, ordered that the ship's spars (the horizontal beams on which the sails hang) be cut down. If the anchors broke, the ship would be driven ashore to almost certain death. The captain of the *Release* did all he could to save his ship: He put three anchors out and cut down the spars, hoping to ride out the tempest, but one anchor broke and the ship was driven to the cliff at White Bear Island. The ship began to leak, and the women and children in the hold noticed the water rising. The *Evening Telegram*, November 12, reported the story in Eli King's own words:

"I went down," said Eli King, a fisherman on the boat, "to see my wife, and she asked me where we were going?"

"I said, we were driving on the cliff!"

"I know where I am going," she replied, "I'm not afraid to die."

"That is good, I said again, and went up to try and get a line ashore for there were some men on the cliff. I wanted to save my eldest boy; I tried hard to save him though I should be drowned myself; but I could not throw the rope so far. I slewed round then to take him in my arms, but he was gone. The sea made a clean breach over the schooner and all in the hold, and my wife and two boys were lost. I don't know how I got ashore. I felt myself in the water and then I felt myself on top of the cliff and that is all I know."

He described the ship's fate: "She went to pieces awful quick, sir. She just bumped three times and then broke up so quick you could not take a hatchet and break up a barrel of flour quicker. Forty-five souls were lost out of eighty; and all I had in the world went down with her."[359]

The ferocity of the storm was recorded in the September 1932 edition of the *Beaver*. The sea rose as a result of tremendous winds and a tidal wave (storm surge), which caused a great deal of the loss of life and destruction of property: "The whole ocean is in one wild upheaval and a smother of white from the driven spume ... the very air reverberating with the cataclysm of sound."[360]

At Cape Harrigan, the sea spray blew over the cliffs (about 91 metres high).[361] All along the coast from Manaks Island, Ragged Island, and White Bear Island, vessels foundered on the high seas or were driven ashore. In total, 150 ships were wrecked and all their gear and cargoes lost; onshore fishing facilities, including stages, flakes, and stores, were also ruined. The *Twillingate Sun* reported that "at one location, a church has been blown down and houses and oil stores have been shaken to pieces."[362] Even those ships which had sought safety in harbours were not immune to the deadly seas. With double anchors and hawsers ashore to ease the strain upon their cables, many were thrown up on the shore and smashed.

One exception to the destruction was the *Panther*, under Captain Bartlett of Brigus, which survived the gale at Manaks Island. She was driven ashore, but the crew re-floated her and then picked up shipwrecked individuals in the general area. In total, this ship transported 500 men, women, and children who were stranded on the shores of Labrador without food or adequate shelter. The *Panther* was the first to tell the news of the disaster to the rest of Newfoundland.[363] When Captain Bartlett arrived in Bay Roberts (Brigus, by some accounts), he immediately telegraphed his employers, Baine Johnston and Co., who alerted the

Newfoundland government and the Honorable Robert Thorburn of the Opposition, who together orchestrated rescue efforts, sending the SS *Lady Glover*, the SS *Mastiff*, and the SS *Plover* from Harbour Grace and St. John's. Newspaper reports praised Thorburn's efforts. The rescue ships were loaded with medical supplies, food, and clothing for the victims and instructed to transport stranded fishermen and their families back to their homes in Newfoundland.[364]

The generosity of the human spirit also shows through in stories of sacrifice. Other vessels that withstood the storm came to the rescue of the shipwrecked. At Smokey Harbour, the barque *Nellie*, ready for her journey to Lisbon with a full cargo, discarded 6,000 quintals of fish, abandoned her voyage, and took on board hundreds of storm victims, transporting them to Newfoundland. The *Lady Elibank* of London under Captain Lee threw all the fish that was on board into the sea and instead transported 400 shipwrecked fishermen and their families to Harbour Grace.[365] The October 29 *Evening Telegram* printed a letter of thanks to Captain Lee:

> Dear Sir,—We, the undersigned mariners and fishermen just arrived in the barquentine *Lady Elibank* from the coast of Labrador, where we were wrecked and cast shore, many of us in an utterly destitute condition, hereby beg to publicly offer our heartfelt thanks to Captain Lee of the above named ship for his kind treatment of us during the passage homeward; and also to testify to our admiration of his skillful management

and navigation of his ship under circumstances of unusual
difficulty, and the excellent discipline which he maintained,
kindly but firmly, notwithstanding that there were upwards
of two hundred souls on board, in men, women and children.
Although prepared to sail to a foreign market, with his
cargo and ship all ready for sea, he did not hesitate to take
us aboard, both himself and his officers giving up their own
berths for our accommodation. And we fully believe that had
we not fallen in with Captain Lee we should, in all probability
have perished or at least suffered unspeakable hardships.
In consideration of these circumstances we feel that we are
unable to properly express out thanks to Captain Lee and his
efficient officers and crew for the care and attention which we
experienced on board his vessel. But we beg to assure him that
his kindness will be long remembered with heartfelt gratitude
by the undersigned, who beg to subscribe themselves: John
Scully, James Deady (for master and crew, schooner *Kenmore*);
Abraham Ledrew, Samuel Glisby, Mark Forward, John Snow,
Job Harvey, John Farrell, Thomas Wall, John Kennedy, Capt.
H. Thomey, John Thomey, W.A. Strapp, on behalf of two
hundred others too numerous to mention.[366]

1942

Perhaps the strong historical and meteorological connection of
the Moravian missionaries with Labrador brought the Nazis to
Northern Labrador to set up their secret automatic weather
station in World War II.[367] On October 22, 1943, a German
submarine dropped off a crew in a rubber boat to install 10 large

canisters filled with nickel-cadmium and dry-cell high-voltage batteries to record and encode weather data and send it back to stations in Northern Europe.[368] The boat landed on the shores of Martin Bay, just south of Cape Chidley in Northern Labrador. This was the only documented landing of German soldiers in North America. The equipment measured atmospheric pressure, temperature, wind direction, and wind speed. An automated weather station was sophisticated technology, and the Allied forces did not seem to be this far advanced. Subsequent reports show that the weather station transmitted data for only a few days before the frequency was jammed. Perhaps the Allies had discovered it.[369]

1947

On October 22, 1947, a wind storm of epic proportions furiously attacked the east coast of Newfoundland. It wreaked misery and destruction from Pistolet Bay on the northern tip of the island to the Avalon Peninsula in the south. The cost of damages was estimated to be in the hundreds of thousands—a large sum of money, particularly as budgets had been depleted during the war effort. From Main Brook on the Northern Peninsula, the residents sent a telegram: "Seas Mountainous high. Everything gone." Perhaps thinking those words were a bit too extreme, considering there was no loss of life, they quickly followed up with the second, more optimistic, message: "All well."[370]

Down the coast at La Scie on the Baie Verte Peninsula, a lighthouse was lost to the storm, and the community's 16-metre-

BONAVISTA IN HEAVY SEAS, CA. 1940

Photo collected by James Swyers, Ross Abbott collection, Bonavista

long bridge narrowly escaped total destruction.[371] At Badger's
Quay, houses shifted off their foundations, and the force of
the wind pulled fences down. In Doting Cove, near Musgrave
Harbour, outhouses floated around the flooded town.[372]
Bonavista, too, experienced heavy seas, which damaged the
breakwater and swept away the Squarry Head lighthouse,
situated on rocks 15 metres above sea level. Fishermen suffered
heavy losses in fishing stages, boats, and fish catches, but no
loss of life. Fortuitously, the fishing season was over for the
year, and most of the boats were tied up in port.[373]

Residents from King's Cove, Bonavista Bay, reported the
"nor'easter" (a storm from the northeast) was "the worst storm
experienced in the memory of the oldest residents." All the
fishing premises in the community, except a stage belonging to
James Stewart, were ruined. The government wharf survived, but
the Aylward brothers, John Hancock, William Costello, and Leo
Sullivan lost part or all of their fishing premises. These fishermen
struggled to save their catch, much of it cured and put up in
storage, only to watch it wash into the ocean with their stages.
In addition to the loss of fish, vital supplies of hay, salt, and
dog food disappeared in the storm. Losses to these fishermen
would take years to replace.[374] The storm struck Catalina with
tremendous force and caused heavy damage to fishing property.
Larenzo Johnson lost his stage and 60 quintals of fish; Sam
Stead's stage and 50 quintals of fish were washed into the ocean;
Joseph Johnson did not lose his stage, but it was left "in a
precarious position." Roads in the area were impassable.[375]

James Maher of Flatrock had a close call when the force of the
water pulled him in and jammed him between the two boats he
was trying to save, his own and one belonging to his father.
The raging sea spit him back, hurling him violently onto the
rocks, where he received several fractures, but escaped with his
life. The power of the sea at Flatrock was so strong that it threw
a large boulder up near the main road.[376]

Captain Dillon, on the MV *Ferryland*, sailed through wind speeds
of 201 kilometres per hour undamaged. He was in the Gulf

Stream at the time, off the Grand Banks. By comparison, 2010's Hurricane Igor brought winds of 140 kilometres per hour— although over open water, with no obstructions, winds speeds can be higher than those on land.[377]

1999

October 15, 1999, was the day the roof blew off the Aquarena, which was built for the 1977 Canada Summer Games. St. John's residents are used to dealing with wind damage, but this event took everyone by surprise. Hurricane Irene lifted the Aquarena's roof 1 metre off the steel-girdered reinforced wooden building before it crashed back down. The peak of the roof split apart, and debris littered the surface of the Olympic-size swimming pool below. The roof emitted a loud groan as it settled back down, and staff quickly evacuated 200 people from the building. It took four weeks to complete the repairs on the building.[378]

Wind speed in St. John's peaked at 106 kilometres per hour that afternoon. High winds and 10-metre waves forced oil rigs to stop drilling. Across the island, power lines went down, causing widespread power outages in St. John's, Corner Brook, Musgravetown, and the Codroy Valley; wires came off houses, trees were uprooted, and schools were closed. The roof blew off the crab plant in Fogo. Due to the storm and high seas, a large piece of floating dry dock eventually washed ashore in Bay St. George on the west coast. It had separated from the larger unit when the cables connecting it to two Russian vessels, which were towing it from China to the Bahamas, snapped.[379]

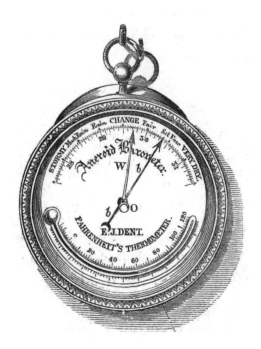

NINETEENTH-CENTURY BAROMETER

NOAA

NOVEMBER

1611

The winter of 1611, the first winter the settlers stayed in Newfoundland under John Guy, was a mild one. Guy wrote from Cupers Cove back home to England:

> The state of the Autumne and Winter was in these parts of New-found-land after this manner. In both the moneths of October and November, there were scarce six days wherin it either freezed or snowed: and that so little, that presently it was thawed and melted with the strength of the Sunne: all the residue of the aforesaid two moneths being both warmer and drier then in England. In December we had sometimes faire weather, some times frost and snow, and sometime open weather and raine: for in latter end it was rainie, and was open weather. All these three moneths the winde was so variable, as it would every fortnight visite all the points of the Compass.[380]

1875

On the night of Monday, November 29, 1875, a furious winter storm with high winds and heavy drifting snow hit as darkness was approaching. It reached the height of its fury at about 8 p.m., sending two ships to their doom. The *Waterwitch*, a 56-tonne, 21-metre-long schooner owned by Charles Bowring of St. John's and built in Trinity, foundered off the coast near Pouch Cove as she battled through the storm on her way to Cupids with winter supplies. Captain Sam Spracklin had 25 crew and passengers, including four women, on board. They had left St. John's at 3 p.m. Although the sky had darkened, and a light snow was falling, there was no sign of the storm they would encounter just outside the Narrows. As the schooner made its way up the coast, conditions worsened. Blinding snow and high winds drove the schooner toward the jagged shoreline. The captain and crew fought valiantly to keep her on course, but it was a losing battle. She was headed for the rocks and sheer cliffs of Horrid Gulch, just outside Pouch Cove. By 8 p.m. the *Waterwitch* had struck the rocks. The captain and two of his men managed to leap from the schooner to a narrow ledge, while other crew members scrambled onto other ledges nearby. With a choice of either being swept off the ledge or climb the vertical cliffs to get help for the stranded crew and the passengers on the ship, the captain and two of his crew chose the latter. They reached the top of the icy cliff and journeyed through blizzard conditions to the north end of Pouch Cove. With great relief, they saw houses, and managed to rouse Eli Langmead with shouts for help. Langmead opened the door to three wet and exhausted men. When he had heard their story, he woke several of his neighbours, and they set off, reaching the site of the disaster at about 1 a.m.

Horrid Gulch is a deep inlet about 2 kilometres northeast of Pouch Cove. The water runs up to the base of steep cliffs, and on one side a narrow ledge stretches along the cliff. There the remaining survivors clung. The rescuers could barely hear their screams and shouts over the noise of wind and waves. In the dark, the only way to ascertain the location of these men so that rescue efforts could begin was to lower a volunteer down over the cliff by rope. Several men hitched the rope around the tree and then held tight to it as Alfred Moores of Pouch Cove stepped forward. Moores went about 156 metres down the cliff in the dark, freezing weather, clinging to the jagged rock. He made three unsuccessful attempts before reaching a ledge over the spot where the survivors were clinging on for dear life. Moores passed a line down to the men, and each of them made it up over approximately 198 metres of steep, jagged cliff face. Thirteen of the 25 people aboard the *Waterwitch* survived the wreck. None of the women, one of whom was the captain's wife, survived.

On the following Wednesday, sleighs from St. John's arrived in Pouch Cove with Mr. Lilly, a Clerk of the Peace, and Mr. Dunphy, from the Poor Office, to help with the aftermath of the catastrophe.[381] The name Alfred Moores was printed in newspapers around North America and in Europe, and the tale of his bravery was told and re-told. He received a silver medal from the Royal Humane Society of England and the citizens of St. John's gave him a silver watch. Five other fishermen also received medals. In Pouch Cove, a plaque marking the beginning of the walking trail to the aptly named Horrid Gulch[382] commemorates the bravery of the fishermen of Pouch Cove who risked their lives to rescue the crew and passengers of the *Waterwitch* that stormy night.

The *Hopewell*, captained by Master Joy, was on her way to Harbour Main when she struck the Biscayne Rock about 1.5 kilometres northeast of Cape St. Francis at 3 a.m. on November 30. Of the eight people on board, one survived. A man by the name of Waugh clung to the rock and was spotted by a passing boat at noon the following day. Those on board lowered a crew to get him. The seas were still running high and the small craft could not get close enough to throw him a line. It took rescuers three attempts and 3.5 hours to get a life preserver to Waugh and transport him back to the steamer. He "soon recovered from the restoratives applied by Captain Blandford of the *Hercules*, whose exertions and those of his crew were above all praise."[383]

1894

A *Daily News* writer described the storm that hit St. Anthony on November 27, 1894, with this succinct statement: "The gale of the 27th ran riot at St. Anthony and did damage enough to delight even the insatiable wind-demon." Three schooners—the *Hannah Jane*, and *Mary Jane*, owned by J. & F. Moore, and the *Porcupine* owned by the Notre Dame Trading Company—were wrecked: "The Methodist church was blown to atoms, even the flooring being torn to pieces by the violence of the Gale ... two stages also felt the effects of the gale, one being quite uncovered." He ends his piece, as if taunting the weather gods: "St. Anthony does not require a repetition of its experience of the 27th."[384]

WEATHER SCIENCE

With so many kilometres of coastline around Newfoundland, breeze invariably blows onshore, even on a fine day. The sea breezes occur particularly in summer due to temperature and pressure differences. The cold air off the ocean rushes into the vacuum left when hot air rises off the land. At night, the opposite can happen: a land breeze, a breeze moving out to sea, is formed.

THE WIND WHISPERER, LAUCHIE MCDOUGALL

The word wind is associated with one name in the minds of many Newfoundlanders: Lauchie McDougall, who lived in the Wreckhouse area, close to the Codroy Valley, on the west coast of Newfoundland and famed for its extremely high winds. In January 1900, for example, a train was blown completely off the track in the Wreckhouse area. The 20-car train, headed for Port aux Basques, was crossing a bridge when five rail cars and the caboose flew off. The caboose landed in a frozen brook on its side.[385] Winds in this area funnel through the Table Mountains and frequently reach triple digit speeds.[386]

McDougall had such an uncanny ability to predict high winds that the Newfoundland Railway paid him to predict the weather. He went outside to check the winds for the company three or four times a day for 30 years, until his death in 1965.[387]

TRADITIONAL WEATHER LORE

If a star is in the crescent of the new moon, it is a sign of bad weather. If the star is anchored away from the crescent, the weather will be fair.

DAMAGE CAUSED BY THE TSUNAMI AT TAYLOR'S BAY, BURIN PENINSULA, NOVEMBER 1929

The Rooms Provincial Archives Division, VA 86-91 / Harris Munden Mosdell

Newfoundland fishermen saw the moon as a ship and the star as a dory.[388]

1929

In the minds of many Newfoundlanders the date of September 18, 1929, is often associated with the unparalled event of a tsunami that hit the south coast of Newfoundland. Although the tsunami was a geological event, weather had a part to play in the disaster. An underwater earthquake rocked the ocean floor 265 kilometres south of the Burin Peninsula at 5:02 p.m. on the evening of Monday, November 18. The quake, although it was felt as far away as New York and Montreal, caused little damage onshore in Newfoundland. The first tremors shook the ground at 5:05 p.m., and the tidal wave that followed at 7:35 p.m. caused destruction and loss of life. The wave started with speeds of 140 kilometres per hour, but then reduced in speed, hitting the shoreline at 40 kilometres per hour. When the wave came in, it swallowed up shore properties and swept them into the sea. For 5 to 10 minutes, sea levels rose 3 to 5 metres above their normal height. Everyone from Burin, on the bottom of the "boot" up to Lamaline 48 kilometres away, felt the earthquake.

Many heartbreaking stories sprang from the disaster. One particularly poignant one told of a mother and three of her children drowned on the first floor of the house while another child on the second floor survived, asleep in bed. In all, 28 people lost their lives, 25 in the tsunami and three afterwards. In St. Lawrence, five-and-a-half-year-old Gus Etchegary was in the kitchen with his sisters. When the quake struck, everything in the kitchen began to shake and a plate on top of a kettle containing the family's supper clattered. Etchegary and his sisters had no idea what was happening, and thought it was the end of the world. The sisters grabbed young Etchegary and ran out the door. As they raced down the hill, they met friends and relatives coming up to gather at the Etchegary home. It was an otherwise beautiful evening, and moonlight lit up the horseshoe of St. Lawrence's harbour. As the townspeople looked out to sea, someone noticed a huge white mass moving toward the harbour. The wave came in with frightening speed and cleaned

out stages and stores. As it sucked back out, it left the harbour
completely dry. About half an hour later, another wave came in.
The Etchegarys and their friends watched in horror, safe from the
deluge, on the hill nearly 150 metres above it all. There was no
loss of life in that community, even though three large waves had
crashed in. The waves cleaned out the harbour and on its return
tossed boats into the woods, 1.5 kilometres away.[389]

A winter storm followed the geological disaster. Just after
daybreak on November 19, the second storm in a few days hit,
dropping temperatures and bringing high winds. Sleet and snow
compounded the misery. The storm caused a tidal surge, which,
combined with a high tide, threw the local population into
panic, as they feared a tsunami reoccurrence.[390]

In the late 1920s, communications between the south coast and
St. John's were rudimentary, with no road connection to the
outside world, and only a single-line telegraph wire connected
communities on the Burin Peninsula to St. John's. The weekend
before the tsunami, a winter storm blew into the region and
knocked out the telegraph wire, cutting the inhabitants off from
any communication except by boat or the still rare wireless
radio.[391] In addition, the tidal wave damaged several local
telegraph stations. It took four days for news of the tsunami
to reach St. John's. The distraught population had been fully
occupied with the search for survivors and tending to the
needs of people left homeless. One ship in Burin harbour had a
wireless radio, but no one knew how to use it. Captain Wesley

B. Kean[392] arrived on the SS *Portia* on November 21 and used his wireless radio to send news of the terrible events.[393]

The full impact of the earthquake was not realized until fishing resumed many months later. It was then discovered that the fishing areas just south of the Burin Peninsula were barren. It was several years before the fishery recovered.[394]

WEATHER SCIENCE

The pressure in the atmosphere is constantly changing. Low pressure and a drop in the barometer indicate poor weather is on the way; high pressure and a high barometric reading indicate fine weather.

TRADITIONAL WEATHER LORE

When the wind backs, and the weather glass falls, then be on your guard against gales and squalls.[395]

A steady, persistent fall in atmospheric pressure is a reliable indication of impending stormy weather. The likelihood of stormy conditions increases when the wind shifts from the west counter-clockwise to an easterly component (northeast or southeast). Such a shift is called backing. In Newfoundland, many winter storms come up from the southwest, as they start to "back" and turn counter-clockwise, the wind then shifts to the east.[396]

SCHOONER ASHORE AT BURGEO

Captain Harry Stone Collection, MHA

DECEMBER

TRADITIONAL WEATHER LORE

Red sky at morning, sailors take warning. Red sky at night, sailor's delight.

Weather systems in the mid-latitudes come predominantly from the west, and if the sky is cloudless in the evening when the sun sets, fine weather will likely follow. Sunlight gets split into the spectrum colours as it hits particles in the normally dusty atmosphere and allows the red rays of the sun through more than the blue. In a high pressure system (good weather) with no clouds to block this phenomenon, a red sky in the evening occurs.[397] A red sky in the east in the morning could mean that the fine weather is exiting and a low pressure system may follow. If it is bright red, it may indicate the atmosphere is heavy with moisture, and it will soon rain.[398]

1612

Crout recorded a red sky on the morning of December 11, 1612:

> ... in the morninge the winde at southwest and be south verie myld weather all the daye thawinge verie much it thawed and uncovered the housses from snowe which had bin Froossen 8 dayes beffore: the sune showing hir selffe abowte 12 of the clock and so contenued allmost untill night this morning was verie close weather with fogg ... the skie in the morning verie

reed ... at the rissing of the sune and verie litell winde mylder day could not be in England in the night after 9 of the clock verie much winde at Southweste with verie much raine all the night: which did consume near all the Froost and snowe which was upon the ground before litell lefte except in some certaine places.[399]

1708

English historian John Oldmixon (1673–1742) described the climate and snow accumulations of Newfoundland in 1708:

> The Climate is very hot in Summer and Cold in Winter; the Snow lies on the Ground 4 or 5 months; and the English in the Northern parts are forc'd to remove from the harbours into the Woods, during that Season, for the convenience of Firing. There they build themselves Cabbins, and burn up all that Part of the Woods where they sit down. The next winter they do the same by another, and so clear 'em as they go.[400]

1779

Thoresby was struck by what he saw on December 17, 1779:

> I was much affected this day whilst passing by the wrecks of five vessels and eleven boats that lay on the beach in Carbonear: This awful scene happened sixteen days ago: language cannot describe the pitiful cries and heart-piercing shrieks! What a spectacle of desolation exhibited to view

when first one and then another of the vessels and boats
clashed against each other, and upon the rocks, till they
became a compleat wreck: and at the same time the poor
men filled with fearful apprehensions that every moment
would be their last: but the blessed God gave them all a kind
reprieve, for some on boards, some in one thing and some
in another, safely got to land.[401]

1874

One hundred years later, the weather had not changed much
from that reported by Thoresby, although the language used to
describe it had. From October to December, Robert Brown, the
Ryan's bookkeeper, described the weather using the adjectives
fine, hard, soft, heavy, disagreeable, dull, or spirited. He calls
the days with frost hard: "it was a fine hard morning"; mild days,
"soft." On December 12, Brown recorded mild winter weather:
"Soft morning, wind W. snow fell during night and roads are
muddy." By December 21 temperatures had dropped:

Wind N.W. in the morning and in the course of the day
it veered around to the N.E. A little snow fell to day. The
ground is now getting quite hard and it is beginning to
look and feel like Winter.[402]

On December 31, Brown wrote an almost apocalyptic description
of the weather. It is easy to imagine him sitting in his frock coat
at his high desk in Bonavista, with a fire burning in the grate,

writing these words: "Fiercely, almost furiously, the year rebels against its doom. The wind is howling dismally, while the sea is turbulent and disturbed."

1876

Brown recorded a heavy snowfall on November 30, 1876, and also reported on several storms that December. After several years of using adjectives like fine, hard, soft, and disagreeable, October 25, 1878 must have brought an unusually nasty storm: "Wind E. weather wretched." On December 30, he recorded a "N.E. 'moderate' breeze with a tremendous sea."[403] One might wonder about his definition of a moderate breeze.

TRADITIONAL WEATHER LORE

If worms are seen on top of the soil in December, the winter will be a mild one.[404]

If the soil is exposed and soft enough for the worms to travel through it in December, it is already a mild winter!

Overly talking about ghosts in a quiet winter's night results in a storm the next day.[405]

1951

West coast Newfoundland communities Stephenville Crossing and Summerside were hammered on December 19, 1951, by a severe winter storm with heavy rain and winds gusting to 180 kilometres

per hour. The rain washed out the railbed and the line was closed
for three days. The train station was flooded and many of the
major roads lay under 1.2 metres of water. The storm blew down
a sawmill at Summerside and 15 telegraph poles. Fishermen lost
boats and gear, and thousands of lobsters washed ashore.[406]

1965

On Christmas Day, 1965, St. John's experienced unseasonably
warm temperatures. The residents enjoyed the mildest
temperature on record since 1944. By Boxing Day, the
temperature had risen to 11°C,[407] and most of the snow in
and around St. John's had disappeared, allowing grass and mud
to peek through. The Avalon Peninsula recorded one of the
warmest temperatures for North America, apart from the most
southerly areas of the United States.[408] Central and western
areas were not so fortunate; they were hit with a snowstorm
that deposited about 23 centimetres of snow in some places
with winds gusting to 121 kilometres per hour. The heavy
snows blocked highways in Central Newfoundland with massive
drifts, and many motorists abandoned their cars wherever
they got stuck. Corner Brook reported snow accumulations of
47 centimetres from Friday to Monday night.[409]

1977

On December 25 and 26, 1977, warm temperatures and rain
melted accumulated snow and caused flood damage to Corner
Brook's streets.[410]

Weather is the cause of up to 75 per cent of major disasters worldwide, according to an estimate by Richard Anthes, president of the University Corporation for Atmospheric Research (Colorado 2008).[411] Canadians, and especially those living in the Atlantic provinces, are continually in the grip of the weather demons, as storms cancel flights, destroy property, cause crop failures, and cost millions of dollars. Newfoundlanders are right to be so mindful of the weather.

NEWFOUNDLAND WEATHER TERMS

Airsome
Cold, fresh, bracing

**Bally-catters
or balli-cutters**
Ice formed by winter spray,
along the shoreline

Brush
A sudden gust of wind
or a spell of wet weather,
usually a snowstorm

Civil
With no wind; still

Conkerbills
Icicles

Duckish
Dark, gloomy, especially at
the end of the day

Fairy squall
A sudden strong gust of
wind on a calm day

Glitter
Ice formed on exposed
objects by freezing rain

In-wind
A wind coming in the bay

Liner
A high wind at the time
of the equinox

Loggy
Heavy with moisture;
oppressively hot

Lun
When winds die down

Mauzy
Damp and warm; muggy

Misky
Misty; wet and foggy

Muggy
Damp, moist, unpleasantly
warm and humid

Norther
A furious cold wind from
the north; a sudden, violent
winter gale from the north

Nor'easter
A wind from the northeast,
usually associated with storms
and violent wet weather

Outwind
A wind going out the bay

Scud
A gust of wind

Scuddies
Sudden gusts of wind; misty,
showery

Screecher
A howling wind

Slob
A slushy dense mass
of ice fragments, snow,
and freezing water

Stun breeze
A sea wind blowing about
20–25 knots

Tickle
A narrow, saltwater strait,
as in an entrance to a harbour
or between islands or other
land masses, often difficult
or treacherous to navigate

Weatherish
Sky clouding over; threatening

Woolly-whipper
A very cold northerly wind
(not specific to Newfoundland
and Labrador)

CONVERSIONS

---

60 fathoms
= 110 metres

4$^{1}/_{2}$ fathoms
= 8.3 metres

40 fathoms
= 73.2 metres

Knot (a measure of speed) × 1.852
= kilometres per hour

Quintal
= 112 pounds or 50.8 kilograms

InHg (a unit of measurement
for barometric pressure) 29.92 inHg
= 1013.25 millibars

Kilopascals (a unit of measurement
for barometric pressure), 101.325 Pa
= 1013.25 millibars

Millibar (a unit of measurement
for barometric pressure), 1013.25
= the standard air pressure in
    millibars at sea level

ENDNOTES

1 Frank McKinney Hubbard, "The Quotations Page," quotationspage.com, 1994–2010, accessed January 2014, http://www.quotationspage.com/quote/26828.html.

2 Roger Price, Meteorological Services of Canada (MSC), Environment Canada, email, August 30, 2010.

3 Ray Bryanton, MSC National Inquiry Response Team, email, June 19, 2013.

4 David Phillips, "Climate and Its Present and Future Impacts on Newfoundland and Labrador," in *Climate and Weather of Newfoundland and Labrador*, ed. Alexander Robertson, Stuart Porter, and George Brodie (St. John's, NL: Creative, 1993), 2–4.

5 "Little Ice Age," NASA, Earth Observatory, accessed May 2014, http://earthobservatory.nasa.gov/Glossary/?mode=alpha&seg=l&segend=n.

6 T.H. Breen, "George Donne's 'Virginia Reviewed'; A 1638 Plan to Reform Colonial Society," *William & Mary Quarterly* 30.3 (1973), 463.

7 Karen Ordahl Kupperman, "The Puzzle of the American Climate in the Early Colonial Period," *American Historical Review* 87.5 (1982), 1286.

8 John Ward Dean, *Capt. John Mason, the founder of New Hampshire: Including his tract on Newfoundland, 1620; the American charters in which he was a grantee; with letters and other historical documents. Together with a memoir by Charles Wesley Tuttle, Ph.D 1815-1902* (Boston: The Prince Society, 1887), 148.

9 Richard Whitbourne, *A discourse and discouery of Nevv-found-land: with many reasons to prooue how worthy and beneficiall a plantation may there be made, after a better manner than it is is, Together with the laying open of certaine enormities and abuses committed by some that trade to that Countrey, and the meanes Laide downe for reformation thereof, Written by Captaine Richard Whitbourne of Exmouth, in the County of Deuon, and Published by Authority. as also a Louing Inuitation: And Likewise the Copies of Certaine Letters Sent from that Countrey; which are Printed in the Latter Part of this Booke* (London: Felix Kyngston, 1620), 55.

10 Edward Wynne, Correspondence from August 17, 1622, accessed February 2014, http://www.heritage.nf.ca/avalon/history/documents/letter_08.html.

11 Sir George Calvert, Lord Baltimore, Correspondence from August 19, 1629, accessed February 2014, http://www.heritage.nf.ca/avalon/history/documents/letter_14.html.

12 Samuel Purchas, *Purchas his pilgrim: or Relations of the world and the religions observed in all ages and places discovered, from the creation unto this present. In foure parts. This first containeth a theologicall and* geological historie of Asia, Africa and America, with the ilands adjacent. Declaring the ancient religions before the floud ... With briefe descriptions of the countries, nations, states, discoveries ... (London: William Stansby for Henrie Fetherstone, 1614), 734.

13 Dean, *Capt. John Mason*, 160.

14 John Mason, *A Briefe Discourse of the New-Found-Land* (Edinburgh: Andro Hart, 1620).

15 Whitbourne, *A discourse and discouery*, 57.

16 Rodney Barney, meteorologist, Environment Canada, email, September 2013.

17 Kupperman, "The Puzzle of the American Climate," 1263.

18 Joel Finnis, Department of Geography, Memorial University of Newfoundland, email, May 2013.

19 Joanna Gyory, Arthur J. Mariano, and Edward H. Ryan, "The Gulf Stream," accessed February 2014, http://oceancurrents.rsmas.miami.edu/atlantic/gulf-stream.html.

20 Benjamin Franklin, *Maritime observations: in a letter from Doctor Franklin, to Mr. Alphonsus Le Roy, member of several academies, at Paris*, in Early American Imprints, Series 1, no. 44888, 314, accessed February 2014, http://qe2a-proxy.mun.ca/login?url=http://opac.newsbank.com/select/evans/44888.

21 David B. Quinn, Alison M. Quinn, and Susan Hillier, eds., *Newfoundland from*

Fishery to Colony. Northwest Passage Searches, in New American World: A Documentary History of North America to 1612, v. 4 (New York: Arno Press and Hector Bye, Inc., 1979), 22.

22 David W. Phillips, *The Climates of Canada* (Ottawa: Environment Canada, 1990), 41.

23 C. Donald Ahrens, Peter Lawrence Jackson, Christine E.J. Jackson, and Christine E.O. Jackson, *Meteorology Today: An Introduction to Weather, Climate, and the Environment*, 1st Canadian ed. (Toronto: Nelson Education Ltd., 2012), 135.

24 C. Donald Ahrens, *Meteorology Today: An Introduction to Weather, Climate, and the Environment*, 10th ed. (Scarborough: Cengage Learning, 2013), 125.

25 Finnis, email, May 2013.

26 Barney, email, September 2013.

27 John Guy, correspondence from May 11, 1611, Journal entries and letter, accessed February 2014, http://www.baccalieudigs.ca/default.php.

28 Samuel Purchas, *Hakluytus posthumus: The Navigations, Voyages, Traffiques, and Discoveries, of the English Nation, to Newfoundland, to the Isles of Ramea and the Isles of Assumption Other wise Called Natiscotec* (Glasgow: J. MacLehose and Sons, 1905), 416.

29 "The International Fishery of the 16th Century," The Newfoundland and Labrador Heritage website project, accessed May 2014, http://www.heritage.nf.ca/exploration/16fishery.html.

30 Jean Baudoin, "Journal of Abbe Baudoin: Diary of a Journey with M. d'Iberville from France to Acadia and from Acadia to Newfoundland," translated by H. Bedford-Jones, St. John's *Daily News*, March 9–15, 1923.

31 Laurence Coughlan, *An Account of the Work of God, in Newfoundland, North America: In a Series of Letters, to which are Prefixed a Few Choice Experiences; some of which were Taken from the Lips of Persons, Who Died Triumphantly in the Faith; to which are Added, some Excellent Sentiments, Extracted from the Writings of an Eminent Divine* (London: W. Gilbert, 1776).

32 Gaston R. Demarée and Astrid E.J. Ogilvie, "The Moravian Missionaries at the Labrador Coast and Their Centuries-Long Contribution to Instrumental Meteorological Observations," *Climatic Change* 91.3-4 (2008), 443.

33 Cornelia Luedecke, "East Meets West: Meteorological Observations of the Moravians in Greenland and Labrador since the 18th Century," *History of Meteorology 2* (2005), 126–27.

34 William Thoresby, *A Narrative of God's Love to William Thoresby: Including, I. His Conversion to God. II. His Call to the Ministry, and Success in Preaching the Gospel, &c. &c. III. His Voyage to Newfoundland. IV. His Labours these, with an Account of several Particular Conversions, &c. V. His Voyage from Newfoundland to England, &c. VI. The Conclusion* (Leeds: printed by Binns and Brown, 1799), 48.

35 Rev. Phillip Tocque, *Newfoundland, as it was, and as it is in 1877* (Toronto: John B.

McGurn, 1878), 455–56.

36 Bruce Whiffen, "What a Year for Weather: Extreme Heat, Cold, Heavy Snow, Torrential Rain—We Saw It All in 2006," *Telegram*, December 30, 2006.

37 "Notes in Brief," *Daily News*, January 10, 1896.

38 Ibid.

39 "Bob-Slide Nuisance," *Daily News*, January 22, 1896.

40 "Notes on the Storm," *Daily News*, January 25, 1896.

41 "Notes in Brief," *Daily News*, January 25, 1896.

42 "Through the Drifts," *Daily News*, January 27, 1896.

43 "Rain Floods Streets in West of the City," *Evening Telegram*, January 30, 1942.

44 Ibid.

45 "Curtail Public Services as a Result of Damage Done by Flood; Family Flooded Out," *Evening Telegram*, January 31, 1942.

46 "Rain Floods Streets in West of the City," *Evening Telegram*, January 30, 1942.

47 "Curtail Public Services."

48 "Bridge Game [sic] Way," *Evening Telegram*, January 31, 1942.

49 "Council Men Busy," *Evening Telegram*, January 31, 1942.

50 "Heavy Rain Made Streets Dangerous," *Evening Telegram*, January 31, 1942.

51 "St. John's Water Supply [Editorial]," *Evening Telegram*, January 31, 1942.

52 Whiffen, "From East to West and Around Again: But the Wind Is Only One Component of a Storm," *Evening Telegram*, January 18, 1999.

53 R.A. Hornstein, *Weather Facts and Fancies* (Downsview, ON: Atmospheric Environment Service of the Department of Fisheries and the Environment, 1977), 19.

54 Phillips, "Climate and Its Present and Future Impacts," 3.

55 Keith C. Heidorn, "The Weather Doctor," 1998–2012, Weather Almanac, accessed February 2014, http://www. islandnet.com/-see/weather/doctor.htm.

56 "Climate, hourly data," Environment Canada, accessed February 2014, http:// www.climate.weatheroffice.gc.ca/ climateData/hourlydata_e.html?timeframe =1&Prov=NFLD&StationID=6773&hlyRan ge=1953-01-01|2013-07-15&Month=1&Day =1&Year=1965&cmdB1=Go.

57 "Some Weather Observations," Canadian Broadcasting Company, Newfoundland and Labrador, 2014, accessed May 2014, http:// www.cbc.ca/nl/features/juneuary/.

58 "Avalon Digs Out after Paralyzing Storm" and "Fireman's Efforts Thwarted by Ferocious School Fire," *Evening Telegram*, January 11, 1966.

59 "West Coast Escapes Worst of Snow Storm," *Evening Telegram*, January 11, 1966.

60 Department of Environment and Conservation, La Manche Provincial Park, accessed February 2014, http://www.env.gov. nl.ca/env/parks/parks/p_lm/.

61 "Wild Seas Lash Coastal Towns," *Daily News*, January 31, 1966.

62 "La Manche: Before and After," *Daily News*, January 31, 1966.

63 "Storms Batter Eastern Canada," *Evening Telegram*, January 31, 1966.

64 "Fishermen Bear Brunt of Severe Storm," *Evening Telegram*, January 31, 1966.

65 "Storm Lashes Avalon Area," *Evening Telegram*, January 31, 1966.

66 "Wild Seas Lash Coastal Towns."

67 "Learning," Met Office, National Weather Service, accessed February 2014, www.metoffice.gov.uk/learning.

68 Heidorn, "The Weather Doctor."

69 "Daily Data Report for January 1977," Environment Canada, accessed May 2014, http://climate.weather. gc.ca/climateData/dailydata_e.html?t imeframe=2&Prov=NFLD&StationID =6711&hlyRange=1970-10-05|1996-07- 01&Year=1977&Month=1&Day=20.

70 "Fiercest Storm of the Winter; St. John's Battered, Province Buried," *Evening Telegram*, January 22, 1977.

71 Bill Kelly and Elizabeth Haines, "City Core in the Dark, Lines Out Everywhere," *Evening Telegram*, January 22, 1977.

72 Sandra Martland, "Infant with Bad Heart, Bundled Up in Drawer," *Evening Telegram*, January 20, 1977.

73 "Storm Causes Limited Damage to Boats, Gear," *Evening Telegram*, January 22, 1977.

74 Bob Wakeham, "Badger Worst Off of All," *Evening Telegram*, January 22, 1977.

75 Earle McCurdy, "Badger Begins Assessing Damage, Major Clean-up Should Start Soon," *Evening Telegram*, January 24, 1977.

76 Eric Witcher, Memorial University of Newfoundland Folklore and Language Archive (MUNFLA) Survey Card 78-325, 1978.

77 "Northern Cod, a Matter of Survival," The History of Northern Cod Fishery Centre for Distance Learning and Innovation, accessed February 2014, https://www.cdli.ca/cod/history4.htm.

78 Richard Hakluyt, *The Principall Navigations, Voyages, Traffiques & Discoveries of the English Nation, vol. 12. (1577)*, ed. Edmund Goldsmid (Edinburgh, Scotland: E. & G. Goldsmid, 1884).

79 Quinn, Quinn, and Hillier, "Newfoundland from Fishery to Colony," 171.

80 Ibid., 177–78.

81 Sir George Calvert and the Colony of Avalon, Memorial University of Newfoundland and the C.R.B. Foundation, 1997–2012, accessed February 2014, http://www.heritage.nf.ca/exploration/avalon.html.

82 Philip Hiscock, "Munsolved Mysteries," *Gazette*, December 12, 1996.

83 Thoresby, *A Narrative of God's Love*, 57, 60.

84 Demarée and Ogilvie, "The Moravian Missionaries," 444.

85 Edwin Noftle, MUNFLA Survey Card 78-207, 1978.

86 Finnis, email, May 2013.

87 Paul O'Neill, interview, April 2010.

88 "Avalanches," Department of Natural Resources, Government of Newfoundland and Labrador, updated October 2012, accessed February 2014, http://www.nr.gov.nl.ca/nr/mines/outreach/disasters/avalanches/.

89 "The Battery, A Case Study," Memorial University of Newfoundland, 1996–2000, accessed February 2014, http://www.heritage.nf.ca/environment/battery.html.

90 "Canada's Wind Chill Index," Environment Canada, updated July 16, 2012, accessed February 2014, http://www.ec.gc.ca/meteo-weather/default.asp?lang=En&n=5FBF816A-1#Wind%20Chill%20Hazards.

91 "Warship and Freighter Lost near St. Lawrence," *Evening Telegram*, February 24, 1942.

92 Cassie Brown, *Standing into Danger*, 1st ed. (Garden City, NY: Doubleday, 1979), 24.

93 Ibid., 27

94 Ibid., 132.

95 Gus Etchegary, interview, April 2013.

96 Brown, *Standing into Danger*, 204.

97 "Eight Men of Lawn Rescue Crew of *Pollux*," *Evening Telegram*, February 26, 1942.

98 *Dictionary of American Fighting Ships* (Washington, DC: United States Government, Department of the Navy, 2004), s.v. *"Pollux,"* accessed February 2014, www.history.navy.mil/danfs/p9/Pollux-ii.htm.

99 "Lanier Phillips: American Hero," The United States Navy Memorial, accessed February 2014, http://www.navymemorial.org/newsroom/news/lanier-phillips-american-hero.

100 "Dead Reckoning: The *Pollux-Truxtun* Disaster," Maritime History Archive, Memorial University of Newfoundland (MHA), accessed February 2014, http://www.mun.ca/mha/polluxtruxtun/index.php.

101 *Encyclopedia of Newfoundland and Labrador* (*ENL*), s.v. "Gaff Topsails."

102 Alex Smith, "Recalls Old Time Winters in Newfoundland," *Evening Telegram*, January 24, 1979.

103 "Cross Country Trains Held Up by Snow Storms," *Evening Telegram*, February 23, 1942.

104 Mary Thorne, interview, February 2013.

105 Charlie Power, interview, August 2013.

106 "Snow Storm Wreaks Havoc," *Daily News*, February 17, 1959.

107 "Mayor Declares City Emergency," *Daily News*, February 18, 1959.

108 "Car Stranded, Occupant Dead," *Evening Telegram*, February 17, 1959.

109 Max Keeping, "Battery Scene of Disaster; Nine Rescued," *Evening Telegram*, February 17, 1959.

110 Ibid.

111 "Please Wait, Mayor Urges," *Evening Telegram*, February 17, 1959.

112 Whiffen, "Wild Is the Wind: What Would You Say Were the Biggest Climatic Events in Newfoundland and Labrador during 1999? Here Are the Top 10 Picks from a Meteorologist's Viewpoint," *Telegram*, December 29, 1999.

113 The Royal Commission on the Ocean Ranger Marine Disaster (Canada), *Report One: The Loss of the Semisubmersible Drill Rig Ocean Ranger and its Crew*, (Ottawa: Royal Commission on the Ocean Ranger Marine Disaster, 1984), 55–56.

114 Terry Roberts, "A Flood of Memories: For One Reporter, the Badger Disaster Really Hit Home," *Western Star*, February 22, 2004.

115 Roberts, "One Year Later, Badger

Residents Remember Devastating Flood," *Western Star*, February 16, 2004.

116 Lorna Nolan, MUNFLA Survey Card 73-171, 1973.

117 Rachel Sawaya, "Science and Nature: How to Forecast Rain and Storms," accessed February 2014, http://suite101. com/article/how-to-forecast-rain-and-storms-a201818 (site discontinued).

118 Barney, September 2013.

119 Quinn, Quinn, and Hillier, "Newfoundland from Fishery to Colony," 174.

120 Noftle, MUNFLA Survey Card 78-207.

121 Barney, email, September 2013.

122 Thoresby, *A Narrative of God's Love*, 45-46.

123 Ibid., 71.

124 Ibid., 78.

125 Phillips, "Climate and Its Present and Future Impacts," 4.

126 "Meteorological Observatory," *Daily News*, September 28, 1898.

127 "A Correction," *Daily News*, September 29, 1898.

128 "Avalanches."

129 *ENL*, s.v. "Southern Cross."

130 Shannon Ryan, *The Ice Hunters: A History of Newfoundland Sealing to 1914* (St. John's, NL: Breakwater, 1994), 311.

131 "The 1914 Sealing Disaster," Memorial University of Newfoundland, 1996–2000, accessed February 2014, http://www. heritage.nf.ca/law/sealing_disaster.html.

132 Mildred Winsor, "Death among the Frozen Ice Floes (A Tragedy at the Seal Fishery)," *Beacon*, February 6, 1985.

133 Morgan MacDonald, "Newfoundland Sealers in Elliston," 2014, accessed April 2014, http://www.morgansculpt.ca/ portfolio/newfoundland-sealers-in-elliston-bonavista-peninsula/.

134 William Tucker, interview, February 2013.

135 Ibid.

136 "Storm Supplement," *Daily News*, March 3, 1958.

137 "Candles for Sale," *Daily News*, March 4, 1958.

138 "Worst Storm in over Thirty Years," *Daily News*, March 3, 1958.

139 Ibid.

140 "Storm Sidelights," *Daily News*, March 4, 1958.

141 "Bell Island without Water. Round the Clock Patrols On," *Evening Telegram*, March 3, 1958.

142 "Conception Bay Power Out ... Week of Tough Work Ahead," *Evening Telegram*, March 3, 1958.

143 Finnis, email, May 2013.

144 O'Neill, interview, 2010.

145 Michelle Osmond, "Red Sky at Night—the Facts and Fiction of Our Weather Folklore," *Downhomer* (November 2004), 43.

146 "Millions in Damage in Eastern Nfld. Following Wind Storm, Sea Surge," *Western Star*, March 18, 2005.

147 Tara Bradbury, "Damage Likely in the Millions," *Evening Telegram*, March 18, 2005.

148 Quinn, Quinn, and Hillier, "Newfoundland from Fishery to Colony," 176.

149 Noftle, MUNFLA Survey Card 78-207.

150 Daniela Guedj and Abraham Weinberger, "Effect of Weather Conditions on Rheumatic Patients," *Annals of Rheumatic Diseases* 49 (1990), 158–59.

151 Thoresby, *A Narrative of God's Love*, 89.

152 Richard Inwards, *Weather Lore* (London: Rider and Company, 1950), 52.

153 Joseph R. Smallwood, "The Snowstorm of March 24, 1939," Transcript of March 24, 1919, radio program *The Barrelman*, reprinted in *The Greenspond Letter* 1.1 (1994), 2.

154 "Train's Snow Bound," *Evening Telegram*, April 8, 1907.

155 "Last Night's Storm: The Worst for Forty Years," *Evening Telegram*, April 8, 1907.

156 Robert A. Cotton, Diaries of Reverend Robert Samuel Smith, NL GenWeb, 2004, accessed February 2014, http://nl.canadagenweb.org/nd_diary1.htm.

157 Gary L. Saunders, *So Much Weather* (Halifax: Nimbus, 2002), 22.

158 Canadian Disaster Database, Public Service Canada, Government of Canada, accessed February 2014, http://cdd.publicsafety.gc.ca/dtpg-eng.aspx?cultureCode=en-Ca&provinces=5&eventTypes=%27FL%27%2c%27SW%27&eventStartDate=%2719840101%27%2%27198412 31%27&normalizedCostYear=1&dynamic=false&eventId=971.

159 Quinn, Quinn, and Hillier, "Newfoundland from Fishery to Colony," 177.

160 Thoresby, *A Narrative of God's Love*, 55.

161 Demarée and Ogilvie, "The Moravian Missionaries," 444.

162 Whiffen, "Fog, Thick Fog, and Pea Soup," *Telegram*, April 13, 1998.

163 Whiffen, "How We Weathered 2001: This Year We Had It All—Record Snowfalls, Flood Damage, Thunder Storms, Intense Blizzards, Hot Flashes and Tragic Losses of Life at Sea. Following Is a List of the Top 10 Weather Stories of the Year," *Telegram*, December 30, 2001.

164 Witcher, MUNFLA Survey Card 78-325.

165 Power, interview, August 2013.

166 "Clearing the Track," *Harper's Weekly*, March 27, 1880.

167 Gerhard Schott, "Die Nebel der Neufundland Baenke," *Die Annalen der Hydrographie & Maritimen Meteorologie* (September 1897), 390.

168 Tucker, interview, January 2013.

169 Whiffen, "One for the Books: The 2000–2001 Winter in St. John's, Newfoundland," *Weatherwise* 57.3 (2004), 48–51.

170 Whiffen, "Oh, Oh, Oh, a Winter of Woe," *Telegram*, April 4, 2001.

171 Nolan, MUNFLA Survey Card 73-171.

172 Barney, email, September 2013.

173 Edward Hayes, *Sir Humphrey Gilbert's Voyage to Newfoundland*, accessed January 2014, ebook #3338, http://www.gutenberg.org/files/3338/3338-h/3338-h.htm.

174 Hakluyt, *The Principall Navigations*.

175 Lester Diaries, 1761–1802, Lester-Garland Records, MHA, accessed February 2014, http://collections.mun.ca/cdm4/description.php?phpReturn=typeListing.php&id=65.

176 Thoresby, *A Narrative of God's Love*, 94.

177 Ibid., 94.

178 W.J. Humphreys, *Weather Proverbs and Paradoxes* (Baltimore: Williams & Wilkins Company, 1923), 73.

179 Ibid., 72.

180 Demarée and Ogilvie, "The Moravian Missionaries," 444.

181 James Ryan Ltd. Diaries, Trinity, Newfoundland Fonds, MHA Coll-011, http://www.mun.ca/mha/viewresults.php?Accession_No=mha00000370.

182 Ryan Diaries, Bonavista, June 2, 1874, MHA 00000370, accessed February 2014, http://collections.mun.ca.

183 Ibid., June 14–15, 1876.

184 Ibid., June 6, 1888.

185 Whiffen, "Flash Chance," *Telegram*, July 19, 1999.

186 Patrick Coish, interview, March 3, 2012.

187 Stephanie Paul, "100th Anniversary of Fire Patrol," Government of Newfoundland and Labrador, Department of Natural Resources, updated August 5, 2013, accessed August 2013, http://www.nr.gov.nl.ca/nr/forestry/fires/protect_centre/anniversary.html.

188 "Case Study: The Bonavista North Fire," in *7 Steps to Assess Climate Change Vulnerability in Your Community*, 6: 22, accessed April 2014, http://www.env.gov.nl.ca/env/climate_change/vultool/pdf/chapter6.pdf.

189 Bob Hyslop, *Journey through Time: Clarenville, Hub of the East Coast* (Clarenville, NL: The Town of Clarenville, 2001), 109–116.

190 Whiffen, "Trivia Time," *Telegram*, January 19, 1998.

191 Bonnie Belec, "Queen's Visit Makes the Day for Many," *Telegram*, June 25, 1997.

192 Lester Diaries, July 16–17, 1767, Lester-Garland Records, MHA.

193 Demarée and Ogilvie, "The Moravian Missionaries," 444.

194 Noftle, MUNFLA Survey Card 78-207.

195 Barney, email, September 2013.

196 Power, interview, August 2013.

197 Liza Piper, "Backward Seasons and Remarkable Cold: The Weather over Long Reach, New Brunswick, 1812–1821," *Acadiensis* 34.1 (2004), 31.

198 Jelle Zeilinga de Boer and Donald Theodore Sanders, *Volcanoes in Human History: The Far-Reaching Effects of Major Eruptions* (Princeton, NJ: Princeton University Press, 2002), 145.

199 Piper, "Backward Seasons," 44.

200 Richard B. Stothers, "The Great Tambora Eruption in 1815 and Its Aftermath," *Science* 224.4654 (June 15, 1984), 1198.

201 Piper, "Backward Seasons," 54.

202 Ibid., 33.

203 Ibid., 51.

204 William Kelson, Slade and Kelson Diaries, Trinity, 1809–1851, Robert Slade & Company Collection, MHA 00000391, June 3, 1816, accessed February 2014, http://www.mun.ca/mha/viewresults.php?Accession_No=mha00000391.

205 Heidorn, "Eighteen Hundred and Froze to Death, the Year There Was No Summer," The Weather Doctor, 1998–2012, accessed February 2014, http://www.islandnet.com/-see/weather/history/1816.htm.

206 Ibid.

207 "A Hundred Years Ago," *Colonial Commerce* 25.1 (December 31, 1915), 25–27.

208 Documents Relating to Battle Harbour and Greenspond, MHA, MF-0185, Provenance: Sean Cadigan.

209 C. Grant Head, *Eighteenth Century Newfoundland, a Geographer's Perspective* (Toronto: McClelland and Stewart Ltd., 1976), 233.

210 Henry M. Stommel and Elizabeth Stommel, *Volcano Weather: The Story of 1816, the Year without a Summer* (Newport, RI: Seven Seas Press, 1983), 87.

211 John P. Newell, "The Climate of the Labrador Sea in the Spring and Summer of 1816, and Comparisons with Modern Analogues," in *A Year without a Summer? World Climate in 1816*, ed. C.R. Harington (Ottawa: Canadian Museum of Nature, 1988), 245.

212 Correspondence between Governor Francis Pickmore and the Earl of Bathurst, The Rooms Provincial Archives of Newfoundland and Labrador, reel #686, Colonial Office (CO) 194, vol. 57, 66.

213 D.W. Prowse, *A History of Newfoundland* (1895; Portugal Cove-St. Philip's, NL: Boulder Publications, 2002), 406.

214 "A Hundred Years Ago," 27.

215 *ENL*, s.v. "Fires."

216 O'Neill, *The Oldest City: The Story of St. John's, Newfoundland* (1979; Portugal Cove-St. Philip's, NL: Boulder Publications, 2008), 449.

217 Ibid., 446.

218 William Nugent Glascock, *Naval Sketchbook; or, the Service Afloat and Ashore*, Vol. 1 (London: J.L. Cox & Son, Great Queen Street, 1831), 168.

219 Ibid., 172.

220 Prowse, *A History of Newfoundland*, 405.

221 *Dictionary of Newfoundland and Labrador (DNL)*, s.v. "ral."

222 de Boer and Sanders, *Volcanoes in Human History*, 153.

223 "A Hundred Years Ago," 26.

224 Michael Harrington, "Summer of the Great Drought," *Evening Telegram*, August 31, 1987.

225 Ryan Snoddon, "The Importance of Weather History, for the Future," Archives and Science: Connections, Exploring the Sciences through Archives, The Association of Newfoundland and Labrador Archives with The Rooms Provincial Archives of Newfoundland and Labrador, November 18, 2011.

226 "City Lashed by Chain of Heavy Rainshowers Lasting 19 Hours," *Evening Telegram*, July 29, 1946.

227 T.A.D., "Regatta Ripples," *Evening Telegram*, July 29, 1946.

228 "City Lashed by Chain."

229 Finnis, email, May 2013.

230 Heidorn, "The Weather Doctor."

231 "Playgrounds Open in Bannerman and Victoria Parks; Mayor and Other Prominent Citizens Join in Ceremony," *Daily News*, July 9, 1949.

232 "City Playgrounds Are Now Open: Bannerman and Victoria Parks Begin 26th Season," *Evening Telegram*, July 9, 1949.

233 "Severe Damage to Crops on Farms of 8 Northeast States," *Evening Telegram*, July 5, 1949.

234 Nolan, MUNFLA Survey Card 73-171.

235 O'Neill, interview, 2010.

236 Whiffen, "How We Weathered 2001," *Telegram*, December 30, 2001.

237 Jamie Harding, artistic director, Beyond the Overpass Theatre, email, January 2014.

238 Barney, email, September 2013.

239 Moira Baird, "Designing Roads," *Telegram*, August 18, 2007.

240 Finnis, email, May 2013.

241 Whiffen, "Wild is the Wind."

242 Witcher, MUNFLA Survey Card 78-325.

243 Barney, email, January 2014.

244 Hakluyt, *The Principall Navigations*, 363.

245 Demarée and Ogilvie, "The Moravian Missionaries," 443.

246 "The Climate of Nova Scotia: A Meteorological Moment," Environment Canada, updated 2007, accessed September 2013, http://web.archive.org/web/20080526093034/http://atlantic-web1.ns.ec.gc.ca/climatecentre/default.asp?lang=En&n=61405176-1.

247 David Longshore, *Encyclopedia of Hurricanes, Typhoons and Cyclones* (New York: Facts on File, Inc., 2008), 74.

248 A.H. Jackson, *Weird Canadian Weather: Catastrophes, Ice Storms, Floods, Tornadoes, Hurricanes and Tsunamis* (Edmonton, AB: Blue Bike Books, 2009), 80.

249 Heidorn, "The Weather Doctor."

250 "Recent Cyclone on the Fishing Banks;

Newfoundland," *Frank Leslie's Illustrated News* 57 (October 20, 1883), 135, 141.

251 Nolan, MUNFLA Survey Card 73-171.

252 James P. Howley, "Reminiscences of Forty-two Years of Exploration in and about Newfoundland 1890–1896," accessed February 2014, http://collections.mun.ca/cdm4/browse.php?CISOROOT=%2Fhowley.

253 "1889: Latest by Telegraph. From Greenspond. Remarkable Thunderstorm. Enormous Hailstones," *Evening Telegram*, August 15, 1889.

254 Ryan Diaries, Trinity, August 5, 1898.

255 Whiffen, "The August Gale: Knocked-out Communications Contributed to Tragedy," *Telegram*, August 23, 1999.

256 Barbara Dean-Simmons, "Twister Tears through Trinity," *Packet*, September 16, 1996.

257 Sue Hickey, "We're Not in Kansas Anymore," *Advertiser*, August 21, 2007.

258 "Storm Impact and Summary, A Climatology of Hurricanes for Eastern Canada, Environment Canada," Environment Canada, accessed February 2014, http://www.ec.gc.ca/hurricane/default.asp?lang=en&n=F77DA7D7-1.

259 John Macleod, senior government archivist, Nova Scotia Archives, email, January 2014.

260 "The August Gales," Fisheries Museum of the Atlantic, Nova Scotia Museum, 2013, accessed February 2014, http://

fisheriesmuseum.novascotia.ca/educational-resources/august-gales.

261 "Disastrous Toll of the Storm," *Evening Telegram*, August 27, 1927.

262 Robert Parsons, "Red Harbour: Death on the Cross Trees," *Newfoundland Quarterly* 88.4 (1994), 25–26.

263 "Many Deaths Caused by Storm," *Evening Telegram*, August 31, 1935.

264 "Sunday's Wind Storm," *Daily News*, August 26, 1935.

265 "Storm Impact and Summary."

266 Kim Todd, "The August Gale of '35, A Survivor's Account," *Downhomer* 15.3 (2002), 25–26.

267 "Widespread Damage by Storm," *Evening Telegram*, August 27, 1935.

268 "Much Damage Done [by] Sunday's Wind Storm," *Daily News*, August 26, 1935.

269 Whiffen, "Wild Is the Wind."

270 Finnis, email, May 2013.

271 Earl B. Shaw, "The Newfoundland Forest Fire of August 1935," *American Meteorological Societies Journals Online* 65.5 (May 1936), 171.

272 Canadian Disaster Database, Public Safety Canada, Government of Canada, updated September 13, 2013, accessed February 2014, http://www.publicsafety.gc.ca/cnt/rsrcs/cndn-dsstr-dtbs/index-eng.aspx.

273 Heidorn, "The Weather Doctor."

274 Whiffen, "Top 10 Weather Events of 2002: Province Saw Usual Snow, Sleet and Rain, As Well As Floods and Hurricanes," *Telegram*, December 29, 2002.

275 Shannon Quesnel, "Uncommon Storm Rakes Region," *Beacon*, August 19, 2002.

276 Witcher, MUNFLA Survey Card 78-325.

277 Quinn, Quinn, and Hillier, "Newfoundland from Fishery to Colony," 158.

278 Phillips, *Climate and Weather of Newfoundland and Labrador*, 3.

279 Jack Williams, *The AMS Weather Book: The Ultimate Guide to America's Weather* (Chicago: University of Chicago Press and American Meteorology Society, 2009), 7.

280 Whiffen, "Stormy Days in 1775: Perhaps You May Have Some Details on This Hurricane," *Telegram*, January 4, 1999.

281 "Person Lately from Halifax to Cape Cod Reports ...," *Pennsylvania Magazine or American Monthly Magazine* (December 1, 1775), 581.

282 Prowse, *A History of Newfoundland*, 653.

283 Alan Ruffman, "The Multidisciplinary Rediscovery and Tracking of 'The Great Newfoundland and Saint-Pierre et Miquelon Hurricane of September 1775,'" *The Northern Mariner / Le Marin du Nord* 6.3 (1996), 15.

284 Lawrence H. Officer and Samuel H. Williamson, "Measuring Worth," 2011, accessed February 2014, http://www.measuringworth.com/ppoweruk/.

285 Ruffman, "The Multidisciplinary Rediscovery," 17.

286 Prowse, *A History of Newfoundland*, 653.

287 "At St. John's, and other places, in Newfoundland, there arose a tempest of a most particular kind," in *The Annual Register, or a View of the History, Politics, and Literature, for the year 1775* (London: The Sixth Edition, 1801), 157.

288 Anne E. Stevens and Michael Staveley, "The Great Newfoundland Storm of 12 September 1775," *Bulletin of the Seismological Society of America* 81.4 (1991), 1400–02.

289 Rev. Lewis Amadeus Anspach, *A History of the Island of Newfoundland* (London: Printed for the author and sold by T. and J. Allman; and J.M. Richardson, 1819), 299–300.

290 Olaf Uwe Janzen, "Newfoundland and British Maritime Strategy during the American Revolution" (PhD dissertation, Queen's University, 1983), 145.

291 L.E.F. English, "Stories of Newfoundland," *Maritime Advocate and Busy East* 1.41 (1950), 12.

292 Dale Jarvis, "The Hollies and the Hurricane—A Chilly Tale of the Mass Grave in Northern Bay Sands," *Downhomer* 16.1 (2003), 70–72.

293 "The Great Gale September 12, 1775,"

The Rooms Provincial Archives Division, MG1000–2000 series box 18, file no. 35.

294 David Wegenast, "Ghosts at Northern Bay," *Decks Awash* 9.6 (1980), 6.

295 George Cartwright, Esq., *A Journal of transactions and events, during a residence of nearly sixteen years on the coast of Labrador; containing many interesting particulars, both of the country and its inhabitants, not hitherto known*, Vol. II (London: G.G. J. and J. Robinson, Paternoster-row and J. Stockdale, 1792), 107.

296 Deana Stokes Sullivan, "The Forgotten Storm," *Telegram*, February 10, 2010.

297 Barney, email, September 2013.

298 Ruffman, "The Multidisciplinary Rediscovery," 19.

299 Ibid., 20.

300 Ibid., 12.

301 Finnis, email, May 2013.

302 C.B.Henn, "Witterungsbeobachtungen, angestellt in Okak auf der Kueste Labrador," *Akademiia Nauk*, SSSR: Bulletin Scientifique 5 (1839), 142–53.

303 "The Storm," *Times and General Commercial Gazette*, September 23, 1846.

304 Ibid.

305 Ibid.

306 "Dreadful Gale!" *Patriot and Terra-Nova Herald*, September 30, 1846.

307 "The Late Tempest," *Royal Gazette*, September 1846.

308 "To the Editor of the *Newfoundlander*," *Newfoundlander*, September 24, 1846.

309 "Dreadful Gale!"

310 "Terrific Hurricane," *Royal Gazette*, September 29, 1846.

311 Ibid.

312 Ibid.

313 Daniel Vickers, *Farmers and Fishermen: Two Centuries of Work in Essex County, Massachusetts 1630–1850* (Chapel Hill, NC: University of North Carolina Press, 1994), 285–86.

314 "Local Happenings," *Evening Telegram*, September 20, 1907.

315 "Postal Telegraphs," *Evening Telegram*, September 21, 1907.

316 "Echoes of the Storm!," *Evening Telegram*, September 23, 1907.

317 "Schr. *Poppy* Rode out the Storm," *Evening Telegram*, September 23, 1907.

318 "Work of the Storm," *Daily News*, September 21, 1907.

319 Benson Hewitt, "The View from Fogo Island," *Pilot*, December 14, 2011.

320 "Lost with All Hands!," *Evening Telegram*, September 23, 1907.

321 "Harbour Grace Notes," *Evening Telegram*, September 21, 1907.

322 "Harbour Grace Notes," *Evening Telegram*, September 23, 1907.

323 "Local Happenings," *Evening Telegram*, September 20, 1907.

324 "Tragedies of the Storm!," *Daily News*, September 23, 1907.

325 "The Big Storm: Toll of the Sea," *Evening Telegram*, September 26, 1916.

326 "City Storm Swept!" *Evening Telegram*, September 25, 1916.

327 Ibid.

328 "Factory and Contents Destroyed," *Evening Telegram*, September 28, 1916.

329 "The Work of the Storm," *Evening Telegram*, September 27, 1916.

330 "Storm Plays Great Havoc at Harbour Grace," *Evening Telegram*, September 25, 1916.

331 "Bay de Verde Notes," *Evening Telegram*, September 29, 1916.

332 "*Portia*'s Stormy Trip," *Evening Telegram*, September 26, 1916.

333 "The Missing *Annie*," *Evening Telegram*, September 28, 1916.

334 "*Portia*'s Stormy Trip."

335 "No Tidings of Crew of Ill-Fated *Bonnie Lass*," *Evening Telegram*, September 28, 1916.

336 "The Work of the Storm."

337 "S.S. *Viking* Got Full Force of the Wind," *Evening Telegram*, September 27, 1916.

338 Ibid.

339 "From Cape Race," *Evening Telegram*, September 25, 1916.

340 "Storm Impact and Summary."

341 "Vigorous Igor," Environment Canada, updated January 28, 2011, accessed February 2014, https://www.ec.gc.ca/meteo-weather/default.asp?lang=En&n=BDE98E0F-1.

342 Ibid.

343 James McLeod, Daniel MacEachern, and Alisha Morrissey, "Army Called in as Province Recovers from Igor," *Telegram*, September 27, 2010.

344 "Snow Storm Wreaks Havoc," *Daily News*, February 17, 1959.

345 Richard J. Pasch and Todd B. Kimberlain, "Tropical Cyclone Report—Hurricane Igor," National Hurricane Centre, February 15, 2011, accessed February 2014, http://www.nhc.noaa.gov/pdf/TCR-AL112010_Igor.pdf.

346 Environment Canada, Special Weather Summary Message for Newfoundland and Labrador, issued by Environment Canada at 4:27 p.m. NDT Thursday, September 23, 2010, accessed February 2014, http://www.atl.ec.gc.ca/weather/bulletins/nf/20100923185938.txt.en.

347 Whiffen, "Riding the Jet Stream," *Telegram*, February 16, 1998.

348 Whiffen, "One for the Books."

349 Quinn, Quinn, and Hillier, "Newfoundland from Fishery to Colony," 159.

350 Ibid., 160.

351 Thoresby, *A Narrative of God's Love*, 41.

352 Ryan Diaries, Trinity, October 10, 1874.

353 Clara J. Murphy, "Labrador Gale of '85," *Them Days* 11.1 (1985), 21.

354 "To the Rescue! An Incident of the Gale," *Evening Telegram*, October 26, 1885.

355 "Latest from Labrador," *Evening Telegram*, October 27, 1885.

356 Murphy, "Labrador Gale," 16–17.

357 "News by the Nellie," *Twillingate Sun*, October 29, 1885.

358 H.M.S. Cotter, "The Great Labrador Gale, 1885," *Beaver* 263.2 (1932), 81.

359 "The Labrador Disaster," *Evening Telegram*, November 12, 1885.

360 Cotter, "The Great Labrador Gale, 1885," 81.

361 Staff Commander W.F. Maxwell, R.N., *The Newfoundland and Labrador Pilot* (London: Great Britain Hydrographic Department, 1887), 451.

362 "News by the Nellie."

363 Cotter, "The Great Labrador Gale, 1885," 82.

364 "Awful News from Labrador," *Twillingate Sun*, October 29, 1885.

365 Cotter, "The Great Labrador Gale, 1885," 82.

366 "The Gale at Labrador," *Evening Telegram*, October 29, 1885.

367 Correspondence, Ruffman to Rev. F.W. Peacock, March 23, 1982, Centre for Newfoundland Studies, Memorial University of Newfoundland, St. John's, NL.

368 "The Germans Were in Labrador during WWII," *Daily News*, August 3, 1981.

369 W.A.B. Douglas, "Beachhead Labrador," *Quarterly Journal of Military History* 8.2 (1996), 35–37.

370 "Wednesday Night's Storm One of the Worst Felt in Newfoundland," *Evening Telegram*, October 24, 1947.

371 Ibid.

372 "Further Reports of Last Week's Storm," *Evening Telegram*, October 27, 1947.

373 "Storm Havoc at Bonavista," *Evening Telegram*, October 23, 1947.

374 "Further Reports of Last Week's Storm," *Evening Telegram*, October 27, 1947.

375 "Wednesday Night's Storm."

376 Ibid.

377 "Newman's Cove and Birchy Cove Notes; October's Disastrous Storm," *Daily News*, November 20, 1947.

378 Barb Sweet and Theresa Ebden, "What a Blowout! High Winds Wreak Havoc across the Province," *Telegram*, October 16, 1999.

379 "1999-Irene," Environment Canada, updated September 14, 2010, accessed February 2014, http://www.ec.gc.ca/Hurricane/default.asp?lang=En&n=AEBAA8E1.

380 Prowse, *A History of Newfoundland*, 125–26.

381 "Loss of the Schrs. *Hopewell* and *Waterwitch* and Nineteen Lives," *Royal Gazette*, December 7, 1875.

382 *ENL*, s.v. "Waterwitch."

383 Jack Fitzgerald, *Newfoundland Disasters* (St. John's: Jesperson, 1984), 89.

384 "Notes from St. Anthony," *Daily News*, December 3, 1894.

385 Joan Dixon, *Extreme Canadian Weather: Freakish Storms and Unexpected Disasters* (Canmore, AB: Altitude Publishing, 2005), 71–72.

386 Whiffen, "Wreckhouse Winds," *Downhome* 13.1 (2000), 54–55.

387 David Phillips, *Blame It on the Weather: Strange Canadian Weather Facts* (Toronto: Key Porter Books, 1998), 153.

388 Nolan, MUNFLA Survey Card 73-171.

389 Etchegary, interview, April 2013.

390 "The 1929 Magnitude 7.2 'Grand Banks' Earthquake and Tsunami," updated April 26, 2013, accessed February 2014, http://www.earthquakescanada.nrcan.gc.ca/historic-historique/events/19291118-eng.php.

391 Jenny Higgins, "The Tsunami of 1929," accessed February 2014, http://www.heritage.nf.ca/law/tsunami29.html.

392 Ruffman and Violet Hann, "The Newfoundland Tsunami of November 18, 1929: An Examination of the Twenty-eight Deaths of the 'South Coast Disaster,'" *Newfoundland and Labrador Studies* 21.1 (2006), 105.

393 "The 1929 Magnitude 7.2 'Grand Banks' Earthquake and Tsunami."

394 Etchegary, interview, April 2013.

395 Humphreys, *Weather Proverbs and Paradoxes*, 60.

396 Whiffen, "From East to West and Round Again," *Evening Telegram*, January 18, 1999.

397 Hornstein, *Weather Facts and Fancies*, 18.

398 "Everyday Mysteries: Fun Science Facts from the Library of Congress," Library of Congress, 2011, accessed February 2014, http://www.loc.gov/rr/scitech/mysteries/weather-sailor.html.

399 Quinn, Quinn, and Hillier, "Newfoundland from Fishery to Colony," 165.

400 *ENL*, s.v. "Winter-Houses and Winter Migrations."

401 Thoresby, *A Narrative of God's Love*, 104.

402 Ryan Diaries, Trinity, December 12, 1874.

403 Ibid., December 30, 1878.

404 Witcher, MUNFLA Survey Card 78-325.

405 Ibid.

406 Canadian Disaster Database, Public Safety Canada, Government of Canada, updated September 13, 2013, accessed February 2014, http://cdd.publicsafety.gc.ca/dtpg-eng.aspx?cultureCode=en-Ca&provinces=5&eventTypes=%27FL%27&eventStartDate=%2719510101%27%2c%2719511231%27&normalizedCostYear=1&dynamic=false&eventId=679.

407 "Weatherman Was Fickle during Holiday Weekend," *Evening Telegram*, December 28, 1965.

408 "Province Favoured with Warm Christmas Weather," *Daily News*, December 28, 1965.

409 "Weatherman Was Fickle."

410 Canadian Disaster Database, Public Safety Canada, accessed February 2014, http://cdd.publicsafety.gc.ca/dtpg-eng.aspx?cultureCode=en-Ca&provinces=5&eventTypes=%27FL%27&eventStartDate=%2719770101%27%2c%2719771231%27&normalizedCostYear=1&dynamic=false&eventId=860.

411 Williams, *The AMS Weather Book*, vi.

ACKNOWLEDGEMENTS

I would like to acknowledge the wonderful help and support of certain individuals and groups: my family, for their patience and understanding while they waited for a book that seemed would never be finished; Bruce Whiffen; Kristie Hickey, Chris Fogarty, and meteorologist Rodney Barney of Environment Canada; retired meteorologist Charlie Power; the late Paul O'Neill; Craig Purchase of Memorial University's Biology Department; Joel Finnis and Norm Catto of Memorial's Geography Department; Sean Cadigan of Memorial's History Department; historian Maudie Whelan; William Tucker; Gus Etchegary; Rev. Father Philip Melvin; Mary Thorne; Rick Coish; the staff at Memorial University's Queen Elizabeth II Library, in particular Joan Ritcey at the Centre for Newfoundland Studies; Memorial University of Newfoundland Folklore and Language Archive (MUNFLA) and the Maritime History Archive (MHA); Helen Miller and staff at the City of St. John's Archives; the staff at The Rooms Provincial Archives of Newfoundland and Labrador; and John Griffin and the staff at the A.C. Hunter Public Library, St. John's.

ABOUT THE AUTHOR

Sheilah Roberts lives with her husband in Portugal Cove-St. Philip's, Newfoundland and Labrador. In addition to enjoying country life and working in Memorial University's Music Resource Centre, she writes. In 2010, Roberts published *For Maids Who Brew and Bake*, a book about seventeenth-century Newfoundland foodways that was named one of the top five books in the special interest category of Cuisine Canada's National Culinary Book Awards. Sheilah has also written for magazines including *Canadian Women's Studies*, *Downhome*, and *Saltscapes*.